LABORATORY
MANUAL

IN PHYSIOLOGY AND ANATOMY

Third edition revised

LABORATORY MANUAL

IN PHYSIOLOGY AND ANATOMY

WITH STUDY GUIDE QUESTIONS AND PRACTICAL APPLICATIONS

ELLEN E. CHAFFEE, R.N., M.N., M.LITT.

Science Coordinator and

Associate Professor

University of Pittsburgh

School of Nursing

J. B. Lippincott Company

Philadelphia • New York • Toronto

Distributed in Great Britain by
Blackwell Scientific Publications, Oxford, London, and Edinburgh

Printed in the United States of America
ISBN 0-397-47313-3

3 4 5 6 7 8 9

To the Instructor

In general, this manual is designed to aid the student in acquiring a practical knowledge of basic scientific facts and principles underlying normal body structure and function. Complicated exercises and theoretical material have been avoided.

During laboratory periods maximum use is made of visual aids such as motion picture films, colored projection slides, models, etc. A complete listing of equipment and sources of supply may be found in the Appendix of this manual.

Living animals as well as fresh specimens from a local abattoir offer the greatest learning experiences of all. During dissection and demonstration periods the instructor has unlimited opportunities to help students in their development of scientific curiosity, powers of observation, and general initiative. The student learns to work as part of a team and to apply the facts, concepts, and principles that were presented in lecture periods.

While major emphasis is on the normal or average body, it also is believed that practical considerations related to disturbances in function serve to whet the appetite for additional study. A captive audience may be a resentful audience unless the program is geared to basic interests as well as basic needs. All too often students are overwhelmed by a mass of dry facts that must be memorized and then regurgitated at periodic intervals as a prerequisite for passing the course.

This manual offers situation-type questions, at the end of each laboratory unit, for the purpose of focusing student attention on the fact that tools which are mastered in this area of study are to be used rather than memorized as isolated bits of incidental knowledge. Not only do these questions stimulate interest in applications, but they provide an important learning experience in the use of allied reference books. However, since pathology should not overshadow the basic objectives, these sections are very brief and merely indicate future potentials.

Whenever possible, a small but complete reference library should be maintained in the laboratory area. In addition to medical dictionaries and standard science textbooks, it would be helpful to have one or two copies of references in such fields as medicine, surgery, obstetrics, etc. Recent news items, pamphlets, and professional journals also are of value. Students will be encouraged to read from more than one book when provision is made for such activity.

Organization of this manual is on the basis of twenty-four laboratory units; however, it is not the intention of the author that such a division be followed blindly.

Examination of the Contents will show that several units are divided into Parts A and B. Many of these subdivisions deserve a full two-hour laboratory period, but such scheduling is dependent on the number of hours allotted to the course. Such time may vary from 90 to 120 hours or more; consequently, no one manual could possibly fit every program without certain adjustments.

Since many schools operate within the framework of "quarters" and/or "semesters," two possible plans for distribution of laboratory units are included on the following page.

The flexibility of this manual can best be appreciated when the instructor becomes familiar with the contents of each unit. For example, study guide questions and practical applications may be done out of class when necessary to conserve time. Consideration also has been given to time requirements for examinations, and an appropriate reduction in content has been made in the introductory units.

The author is indebted to Virginia G. Braley, R.N., Ph.D., Associate Professor, University of Pittsburgh School of Nursing, Pittsburgh, Pennsylvania for her helpful suggestions during the revision of this manual. The many excellent line drawings are the work of Ellen Cole, Medical Illustrator, of Philadelphia.

Ellen E. Chaffee

Sample Distributions of Laboratory Units.

Plan A: 2 quarters
(24 weeks)

lecture—2 hours
laboratory—2 hours

Week	Lab. Unit
1:	1
2:	2
3:	3
4:	4
5:	5
6:	6
7:	7
8:	8
9:	9, (A)
10:	9, (B)
11:	10
12:	11 & 12, (A)
13:	12, (B) & 13
14:	14
15:	15
16:	16
17:	17
18:	18
19:	19, (A)
20:	19, (B)
21:	20
22:	21
23:	22
24:	23 & 24

Plan B: 1 semester
(16 weeks)

lecture—2 hours
laboratory—4 hours

Week	Lab. Unit
1:	1
	2, (A)
2:	2, (B)
	3
3:	4
	5, (A)
4:	5, (B)
	6, (A)
5:	6, (B)
	7, (A)
6:	7, (B)
	8
7:	9, (A)
	9, (B)
8:	10
	11
9:	12
	13
10:	14
	15
11:	16
	17, (A)
12:	17, (B)
	18
13:	19, (A)
	19, (B)
14:	20
	21
15:	22
	23
16:	24, (A) & (B)
	24

Contents

To the Student

This manual was written for you! It has been carefully designed to help you to better understand the amazing efficiency of the human body. Physiology and Anatomy are two of the most fascinating subjects in the entire curriculum, and you will find that time speeds by much too quickly when laboratory work is in progress.

The benefits that you will derive from these exercises, however, are directly proportional to the amount of energy and enthusiasm that you bring to class. The instructor will point out many new and exciting pathways, but *you* must do the actual exploring. It is here in the laboratory that lecture material "comes to life," but it can be seen only by those who are willing to open their eyes.

There will be numerous opportunities to work with fresh and preserved tissues. Such experiences will enable you to develop a scientific attitude, increase your ability to make accurate observations, and serve as a means of evaluating your skill in utilizing lecture material.

A large number of diagrams have been included to help you during preliminary studies and again during the process of review. When labels are neatly printed at the end of each line it greatly facilitates future use of your manual.

Applications to practical situations, as well as study guide questions, have also been included. These offer many additional opportunities for professional growth and development. Use more than one reference book in seeking answers to these questions, and then record the information in your own words rather than copying verbatim from the text. Such practice in organizing your thoughts will prove valuable in the future when writing examinations.

Above all, learn to do your own work! No one can learn for you, and a willingness to solve problems with one's own ingenuity is particularly welcome in the study of any science.

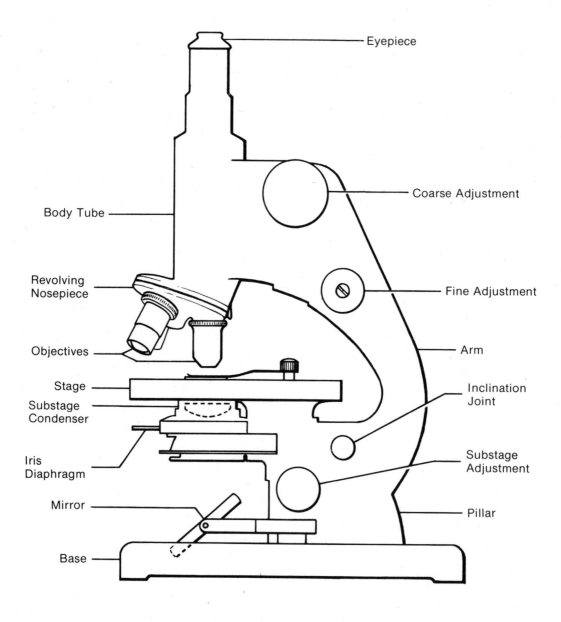

Eyepiece

Coarse Adjustment

Body Tube

Revolving Nosepiece

Fine Adjustment

Objectives

Arm

Stage

Inclination Joint

Substage Condenser

Iris Diaphragm

Substage Adjustment

Mirror

Pillar

Base

Many students have used microscopes during course work in general biology, zoology, etc. The above diagram is included for those individuals who have not had such experience and for those who would benefit from a review. The instructor will demonstrate the proper care and use of this valuable instrument.

LABORATORY 1

Introduction to the Body as a Whole

Demonstration Supplies:
  anatomical wall charts
  dissectible human torso model

Student Supplies:
  live frog for each two students
  dissecting trays and instruments

I. *Demonstration of Manikin.*

The instructor will demonstrate the human torso model while pointing out the various parts and regions of the body. Pay particular attention to the cavities and their contents.

II. *Introduction to Dissection.*
Another way to study body organization is by dissection of living frogs. However, when experimental work is to be done with frogs it is necessary to eliminate all sensations of pain. The instructor will do this by administering anesthetics or by destroying brain tissue (pithing).

Dissection of the frog:

A. Place frog ventral side up in the dissecting tray.

B. With forceps pick up the skin over the pubic region and snip a small "v" shaped incision with the scissors.

C. Insert scissors through this opening and cut through the skin up to the lower jaw. Avoid the superficial blood vessels by cutting slightly to one side of the midline.

D. Notice the blood vessels on the under surface of the skin and in the muscular body wall.

E. Following the same pattern, make an incision through the muscular wall of the abdomen. Avoid blood vessels when possible.

F. The sternum (breastbone) also should be cut in the midline. Keep the point of your scissors close to the under surface of the bone to avoid cutting the heart.

G. Pin frog to the wax in your dissecting tray so that the internal organs are clearly visible. Note the absence of a diaphragm. Occasionally, one or both of the lungs may be distended with air and must be collapsed by a snip with your scissors.

H. Identify the following organs:

adrenal gland
esophagus
fat body
gallbladder
heart
kidney

liver
lung
large intestine
mesentery
pancreas
pharnyx

small intestine
spleen
stomach
testis (eggs and
oviducts in female)
tongue
urinary bladder

I. Upon completion of the dissection, wrap the frog in newspaper and place it in the appropriate waste container.

J. Wash and dry all equipment before returning it to the supply table.

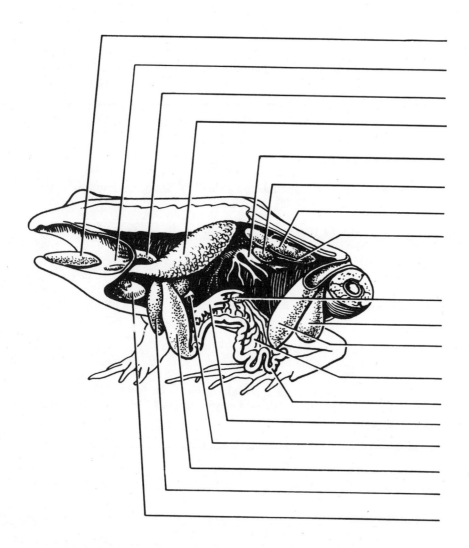

Figure 1.

III. During the remainder of the period examine the torso model, wall charts, and reference books as you complete the following study guide questions:

A. The chest is _____ to the abdomen and _____

to the neck.

B. The foot is _____ to the ankle, but it is _____

to the toes.

C. The kidneys are _____ to the vertebral column.

D. The cranial nerves are _____ to the brain.

E. The _____ surface of the nose is covered with skin, the _____

surface with mucous membrane.

F. The little finger is on the _____ side of the hand, and the

thumb is on the _____ side.

G. The kneecap or patella is on the _____ or _____

aspect of the lower extremity, but the elbow is on the

_____ or _____ aspect of the

upper extremity.

H. A line through the middle of the forehead, tip of the nose, and umbilicus would be in a
_____ plane.

I. A _____ plane would divide the body into front and back portions.

J. If you put your hand flat on the desk, the _____ surface will be touching.

K. When you walk, the _____ surface of your foot touches the floor.

L. The stomach is located in the _____ and _____

_____ regions of the abdomen.

M. To remove your appendix the doctor would make an incision in the _____

_____ region of the abdomen.

N. The urinary bladder is located in the _____ region of the abdomen.

4

IV. *Label the diagrams found on pages 5 and 6.*

Print at the *end* of each label line.

A. *Regions of the body.*   *Anterior aspect.* (page 5).

axillary lumbar, left
brachial mammary
epigastric patellar
femoral pectoral
frontal pubic
hypochondriac, left temporal
hypogastric umbilical
iliac or inguinal, left volar (palmar)

B. *Regions of the body.*   *Posterior aspect.* (page 6).

cervical occipital
deltoid parietal
gluteal popliteal
lumbar region of back sacral
mastoid scapular

A. *Regions of the Body.*   *Anterior Aspect.*

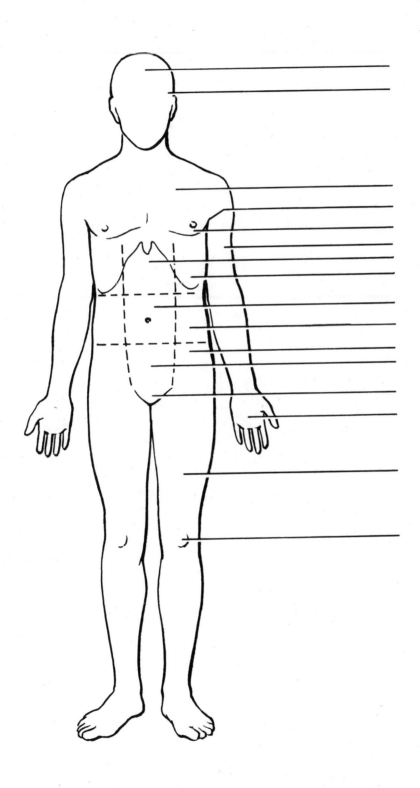

Figure 2.

B. *Regions of the Body.*   *Posterior Aspect.*

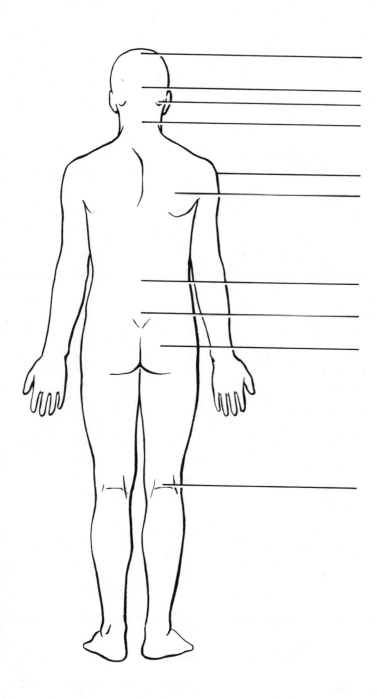

Figure 3.

V. Locate the preceding parts and regions on yourself or your partner.

VI. With the aid of your text, wall charts, and the manikin, list the organs found in each of the following cavities:

CAVITY	ORGANS
Cranial Vertebral	
Thoracic	
Abdominal Pelvic	

LABORATORY 2

Structural Units

Part A — **The Cell**

Demonstration Supplies:
    powdered wood charcoal
    copper sulfate crystals
    potassium permanganate
    sodium chloride
    ammonia or ether
    filter paper
    dried prunes
    distilled water
    glass funnel
    6 beakers
    6 test tubes on rack
    dializing shell and membrane
    gelatin
    egg albumin
    ring stand

Student Supplies:
    1% methylene blue
    95% alcohol
    toothpicks
    glass slides
    blotting paper
    microscopes
    amoeba culture

I. *Examination of Human Cells.* (as found in the mouth)

   A. With the broad end of a toothpick gently scrape the lining of your cheek two or three times to collect some cells from the mucous membrane.

   B. Move the toothpick across a clean glass slide leaving a very thin layer of material.

   C. Allow this slide to air dry.

   D. Now place several drops of 95% alcohol on the slide so that the smear is covered. Allow this preparation to stand for 1 minute.

   E. Shake off the alcohol and carefully blot the slide dry.

   F. Next cover the smear with several drops of 1% methylene blue dye or comparable staining solution. Stain for 1 minute.

   G. Gently rinse slide in some cold water to remove excess dye. Carefully blot the slide dry.

   H. Examine the slide under the microscope. Use both low and high power.

   I. Sketch a few cells just as they appear under your microscope.

   J. Label cell membrane, nucleus, and cytoplasm.

II. *Cell Structure and Function.*

   Briefly summarize the function of each of the following structures, and then label the diagram of a typical cell. (Fig. 4).

   1. cell membrane _____

   2. centrioles _____

3. chromatin _____

4. cytoplasm _____

5. endoplasmic reticulum _____

6. Golgi apparatus _____

7. lysosome _____

8. mitochondrion _____

9. nucleus _____

10. nucleolus _____

11. ribosomes _____

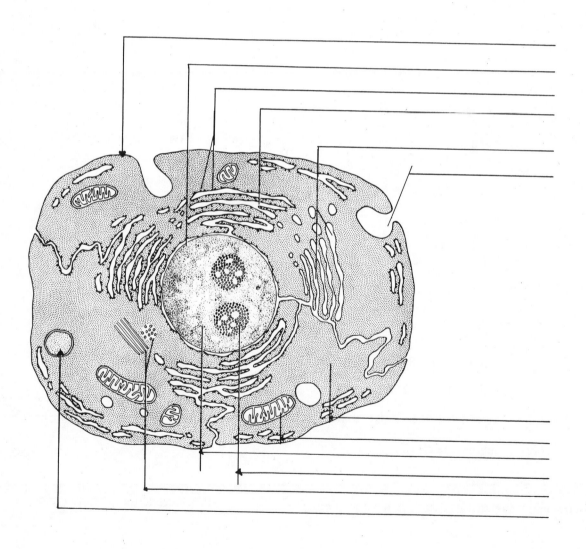

Figure 4.

III. *Comparative Study of the Amoeba.*

A. Place a drop of amoeba culture on a glass slide. Cover with a glass cover slip.

B. Examine under the microscope using low power until an amoeba is located. You may find it helpful to close the iris diaphragm slightly thus reducing the light.

C. Watch the amoeba for a few minutes. Try to identify the following structures:

cell membrane   nucleus
cytoplasm   pseudopodia
food vacuole

D. Pay particular attention to the manner in which the amoeba moves. The white blood cells in your body are capable of similar motion.

IV. *Maintaining Homeostasis in the Human Cell Environment.*

All cells must have a means of receiving nutrients and eliminating waste products. The instructor will present elementary demonstrations of the general processes involved in this exchange. Answer the related questions:

A. *Filtration.*

A mixture of powdered wood charcoal, copper sulfate, and water is poured through filter paper.

1. Which substances pass through the filter paper?

_____

2. Explain: _____

_____

3. Give one example of filtration within your own body:

_____

B. *Diffusion of gases.*

A bottle of ammonia (or ether) is opened at the front of the room.

1. Why may the odor soon be detected in all parts of the room?

_____

2. Where might we find a prominent diffusion of gases taking place in the body?

_____

C. *Diffusion within a liquid.*

A crystal of potassium permanganate is dropped into a beaker of water. Observe this container at intervals during the laboratory period.

1. What happens? _____

   _____

2. Give one example of diffusion through a liquid medium in your own body:

   _____

D. *Diffusion through a colloid.*

Various dyes are placed in stab-depressions in solidified gelatin. At the end of the period examine this material and see what has happened.

1. Explain: _____

   _____

2. The gelatin might be said to resemble what part of a cell?

   _____

E. *Diffusion through a membrane.*

A mixture of water, sodium chloride (a crystalloid) and egg albumin (a colloid) is placed in a dialyzing shell or in a cellophane bag. This preparation then is carefully placed in a beaker of distilled water. Toward the end of the laboratory period the instructor will test the 2 liquids for the presence or absence of sodium chloride and egg albumin. For example, silver nitrate ($AgNO_3$) will react with sodium chloride to form a white precipitate (silver chloride); nitric acid (HNO_3) will coagulate the protein albumin.

1. What was found in the artificial cell? _____

   _____

2. What was in the fluid surrounding the artificial cell?

   _____

3. Explain what has happened: _____

   _____

   _____

4. How can this demonstration be applied to the cells in your own body?

   _____

   _____

F. *Osmosis.*

A dried prune is placed in a beaker of water. Examine it at the end of the period.

1. What has happened to the prune? _____

2. Explain: _____

    _____

3. What has appeared in the water surrounding the prune?

    _____

4. How did it get there? _____

A small bag of dializing membrane or cellophane is filled with a saturated sugar solution. The top is tied securely around a rubber stopper fitted with glass tubing, and all air displaced from the bag. This preparation then is suspended in a beaker of distilled water. (Note: carmine dye may be added to the sugar solution to simulate blood.)

1. Examine the glass tubing at periodic intervals. Record your observations and explain what has happened:

    _____

    _____

2. How could this be applied to your own body?

    _____

G. The following diagrams represent beakers of water with varying concentrations of sodium chloride as indicated. Each beaker contains a human cell with the normal sodium chloride content of 0.9%.

Indicate, by the use of arrows, the direction of movement of water molecules in each case. Also sketch any expected changes in the shape of the cells, and then answer the questions on the following page.

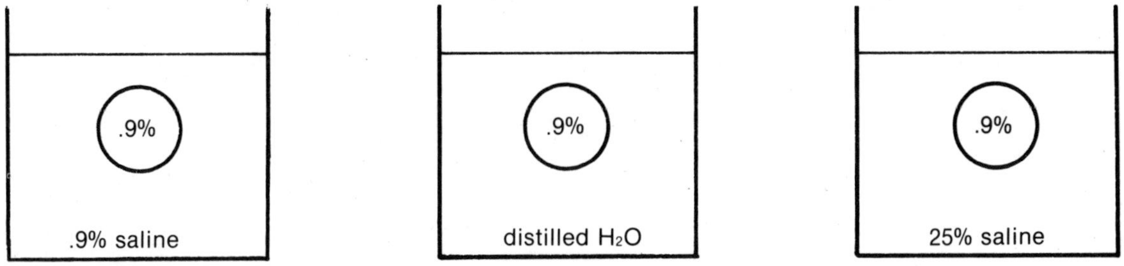

.9% saline   distilled H₂O   25% saline

Figure 5.

14

1. What is the effect of 0.9% saline on the cell? _____

   Explain: _____

2. What is the effect of distilled water on the cell?

   _____

   Explain: _____

   _____

3. What is the effect of 25% saline on the cell? _____

   Explain: _____

   _____

4. What determines the osmotic pressure of any given solution?

   _____

5. A 5% glucose solution is isotonic to cellular fluid, but a 5% saline solution is hypertonic.

   How can this be explained? _____

   _____

V. *Diagram of a Cell and its Blood Supply.*

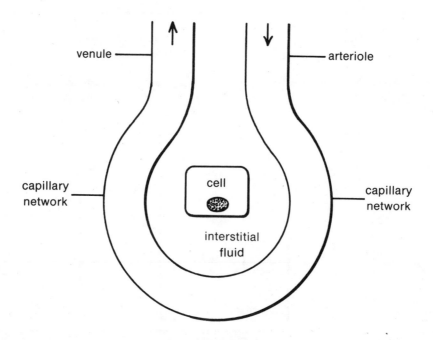

Figure 6.

A. On the above diagram indicate the relative concentrations of nutrients and wastes in the blood vessels. Use "X" to represent foods and "O" for the waste products. Colored pencils may also be used. The space between letters shall represent water molecules.

B. By the use of arrows indicate the pathway taken by foods and wastes in relation to the cell.

C. Compare the two major pressures of the blood as found in the capillary area:

	arterial portion of capillary	venous portion of capillary
Hydrostatic Pressure ("pushing pressure")	_____ mm. Hg	_____ mm. Hg
Colloid Osmotic Pressure ("pulling pressure")	_____ mm. Hg	_____ mm. Hg

D. While keeping the above pressures in mind show, by the use of dotted or broken arrows, the direction of movement of the majority of water molecules. Indicate whether the process is primarily filtration or osmosis.

for example: $\xrightarrow{f}$ or $\xrightarrow{o}$

Part B — **Tissues and Membranes**

Demonstration Supplies:
    fresh chicken leg
    dissecting instruments
    microscopes
    prepared slides of tissues
    delineascope or slide projector (optional)

I. *Demonstration of Fresh Tissues.*

The instructor will demonstrate various tissues in a fresh chicken leg. (Note: save bones for use in Laboratory 4., p. 25.)

II. *Identification of the Primary Tissues.*

From your study of the prepared slides and textbook illustrations of tissues identify the following diagrams:

adipose
cartilage, hyaline
columnar, ciliated
columnar, goblet cells
muscle, cardiac
muscle, skeletal
muscle, visceral

nerve cell (neuron)
osseous
squamous, simple
squamous, stratified (keratinized)
dense fibrous

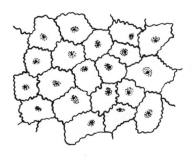

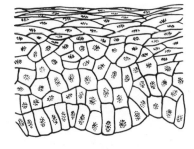

A _____   B _____   C _____

Figure 7.

Figure 7. (cont.)

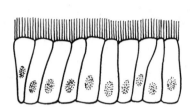

D _____

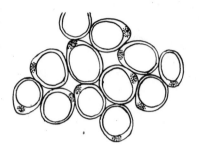

E _____

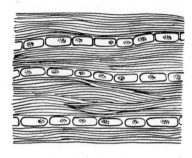

F _____

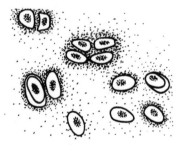

G _____

H _____

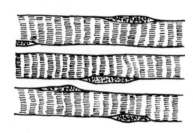

I _____

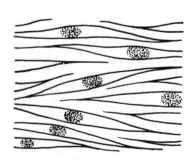

J_____

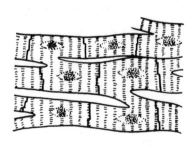

K _____

L _____

III. *Outline Summary of Epithelial and Connective Tissues.*

This table may be used for review purposes.

Name of Tissue	Location in the Body	Function in that Area
A. Epithelial Tissues: 1. simple squamous		
2. simple cuboidal		
3. simple columnar a. goblet cells		
b. ciliated cells		
4. stratified squamous a. keratinized		
b. non-keratinized		
5. transitional		
6. glandular a. exocrine		
b. endocrine		

(continued on page 18)

Name of Tissue	Location in the Body	Function in that Area

B. Connective Tissues:
  1. ordinary loose
    (areolar)		

2. adipose		
3. dense fibrous a. tendons		
b. ligaments		
c. elastic fibers are prominent		
4. cartilage a. hyaline		
b. fibrous		
c. elastic		
5. osseous		
6. hemopoietic		

IV.   The following table may be used for review of the membranes.

Type of Membrane	Combination of What Tissues	Location in the Body	General Function
mucous			
serous			
synovial			
superficial fascia			
deep fascia			
miscellaneous fibrous			

LABORATORY 3

The Integument

Demonstration Supplies:
  anatomical wall chart
  model of human skin
  prepared slides of human skin
  microscopes or projector

Student Supplies:
  blunt and sharp probes
  beakers of hot and cold water
  plain dividers
  glass slides and cover slips
  microscopes
  touch discrimination boxes (coins, cloth, etc.)

I. *A Study of Human Skin.*

   Examine the prepared microscope slides of human skin. Identify the following structures:

        epidermis
            stratum corneum
            stratum germinativum

        dermis
            papillary layer   hair and follicle
            reticular layer   arrectores pilorum
            sebaceous glands   (pilomotor muscle)
            sweat glands

II. *A Study of Human Hair.*

   Place a short strand of hair on a glass slide and secure it with a cover slip. Examine under high and low power with the microscope.

   Identify: root and shaft.

III. *Examine Your Fingernails.*

   A. What and where is the lunula? _____

   _____

   B. What is a common name for eponychium? _____

   C. What can be done to prevent hangnails?

   _____

IV. *The Skin and Interpretation of the Environment.*

   A. *Discrimination of touch sensations.*

A box containing familiar items (coins, cloth, etc.) may be obtained from the supply table. With your eyes closed or with your hands behind your back identify each article as it is handed to you by your partner. Test the sensitivity of each finger and of different parts of your hands.

1. Where is sensation most acute? _____

   _____

2. What specific end organs are stimulated? _____

   _____

B. *Discrimination of other cutaneous sensations.*

1. Have your partner close both eyes, then test the skin on one forearm for perception of heat and cold. Use metal probes which have been standing in beakers of hot and cold water (remove excess water from probe before touching skin).

   Record your results: _____

   _____

2. Also test for discrimination between sharp and dull instruments. Use teasing needles and blunt probes.

   Record your results: _____

   _____

3. Using the plain dividers (or 2 sharp probes) hold the points at varying distances apart. Gently touch the skin of your partner's hand, forearm, arm, nape of neck, etc. See if he or she can tell when *both* points are in contact with the skin.

   Record the range of variation: _____

   What areas of the body were least sensitive? _____

   _____

   What area was the most sensitive? _____

   _____

V. *Application to Practical Situations.*

A. The following questions may be asked by friends who know that you are studying anatomy and physiology. How would you answer them in simple terms?

1. What are freckles? _____

2. Are special freckle-remover creams of any value? _____

   Why? _____

   _____

3. How often should I wash my hair? _____

   _____

4. Why does my sun tan seem to fade in the winter? _____

_____

_____

_____

5. Why does a hypodermic needle hurt me some times but not others? _____

_____

6. My doctor told me that it was dangerous to squeeze pimples and boils. Why? _____

_____

7. What is a blackhead? _____

_____

8. Why do we get "goosepimples" and what makes our hair stand up when we are cold? _____

_____

B. What layer of the skin has no direct blood supply? _____

_____

Why is this fact of special significance to bedridden persons? _____

_____

C. Blood vessels in our skin play an important part in the regulation of body temperature. Explain briefly: _____

_____

_____

_____

1. Other *major* functions of the skin are _____

_____

2. What are the minor functions of the skin? _____

_____

D. Explain the following terms:

1. urticaria _____

_____

2. dermatitis _____

_____

3. abrasion _____

_____

4. diaphoresis _____

_____

LABORATORY 4

Introduction to the Skeletal System

Bone Structure

Demonstration Supplies:
    2 fresh long bones (beef)
        (longitudinal and transverse sections)
    dissecting instruments
    small bone soaked in 10% nitric acid
    small bone subjected to intense heat
        (prepared from chicken legs used in Lab. 2).
    prepared slides of human bone tissue
    microscopes

1. *Macroscopic Structure of Bone.*

    A. The instructor will demonstrate the structure of fresh beef bones which have been specially sectioned.

    B. Examine the small bone which has been immersed in acid and the bone which has been subjected to intense heat.

    C. Answer the following questions:

        1. What is the anatomical term used in reference to the ends of a long bone? _____

        2. What type of tissue covers this area? _____

        3. What is the function of this tissue? _____

        4. What anatomical term refers to the shaft of a long bone? _____

        5. Name the fibrous membrane that covers the shaft. _____

        6. What are the functions of this membrane? _____

        7. Where is compact bone tissue located? _____

        8. Where would you find cancellous bone tissue? _____

9. Give logical reasons for this placement of compact and cancellous bone: _____

_____

_____

10. What blood vessels supply the bone as a whole with nourishment? _____

_____

11. Distinguish between red and yellow bone marrow:

	red marrow	yellow marrow
location:		
composition:		
function:		

12. How can you best describe the bone that has been soaked in acid? _____

Explain: _____

13. What happened to the bone that was subjected to intense heat? _____

Explain: _____

14. What are your conclusions regarding the composition of a functional bone? _____

_____

_____

II. *Microscopic Structure of Bone.*

A. Examine the prepared slides of bone tissue. Use both low and high power magnification.

Draw a simple diagram of one haversian system and label the following structures:

    canaliculi   lacunae

    haversian canal   lamellae

B. Examine the prepared slide of developing bone, cartilaginous type. Under low power magnification identify the hyaline cartilage and preliminary bone tissue. Make a rough sketch of this arrangement.

C. Answer the following questions:

1. Bone building cells are called _____

2. They eventually become imprisoned in the _____

   of the haversian systems and cease to form bone.

3. Following their imprisonment these cells become known as _____

4. How do these trapped but living cells receive food and eliminate waste? _____

   _____

5. What inorganic salts are most abundant in bone tissue? _____

   _____

III. *Application to Practical Situations.*

Henry K. has been admitted to the hospital with a diagnosis of green stick fracture, left femur.

1. How does this differ from an ordinary fracture? _____

   _____

2. What does the diagnosis tell you about his age? _____

   _____

3. Why does the green stick type of fracture occur only in this general age group? _____

   _____

4. In addition to the fracture, x-rays also reveal obvious bands of cartilage near the epiphyses of Henry's femur. What does this indicate? _____

5. What are the functions of the skeleton?

   _____

   _____

6. Which of these functions are most seriously interfered with when a bone is broken? _____

   _____

7. Which functions of the periosteum are of increased importance following a fracture? _____

   _____

8. What foods should Henry eat to build strong bones? _____

   _____

LABORATORY 5

The Axial Skeleton

Part A — **Bones of the Skull**

Demonstration Supplies:
   fetal skull
   dissectible, adult skull

Student Supplies:
   plain adult skulls (hinged mandible and calvarium)

I. *A Study of the Human Skull.*

Examine one of the plain white skulls. With the aid of text and reference books identify the bones and their markings. How many of the skull bones can you locate on yourself?

II. *Demonstration.*

The instructor will meet with small groups and demonstrate the fetal and adult skulls. Be prepared to answer questions on the cranial and facial bones.

III. *Label the diagrams on the following pages.*

PRINT all labels at the *end* of the appropriate line.

   A. *Fetal Skull.*

   Fontanels   *Sutures*

   anterior   coronal
   antero-lateral   sagittal
   posterior
   postero-lateral

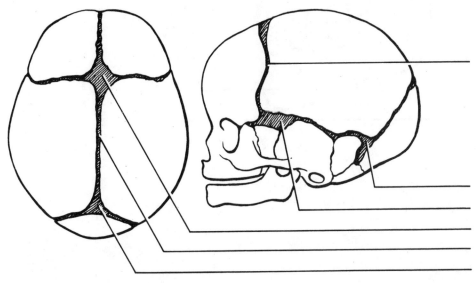

Figure 8.

30

B. *Lateral Aspect of Skull.*

Bones	*Sutures*	*Processes*
frontal	coronal	condyle of mandible
lacrimal	lambdoidal	mastoid
mandible	squamosal	
maxilla		
nasal		
occipital		
parietal		
sphenoid		
temporal		
zygomatic		

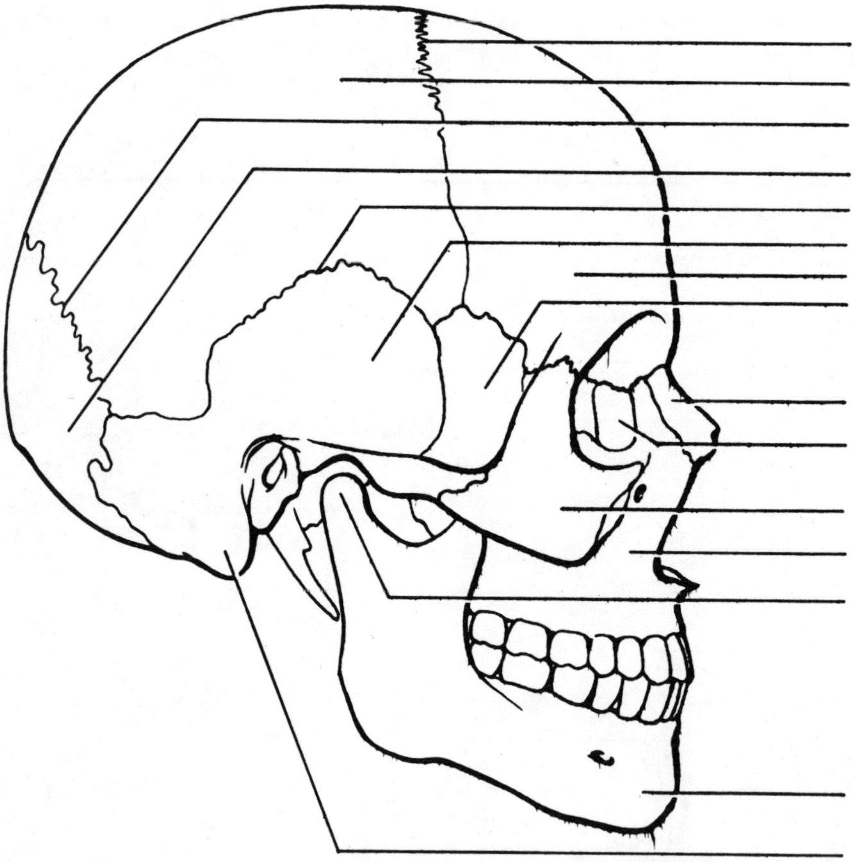

Figure 9.

C. *Anterior Aspect of Skull.*

Bones

frontal
inferior concha
lacrimal
nasal
parietal
sphenoid
temporal
vomer
zygomatic

Processes

alveolar process of mandible
alveolar process of maxilla
mental foramen
perpendicular plate

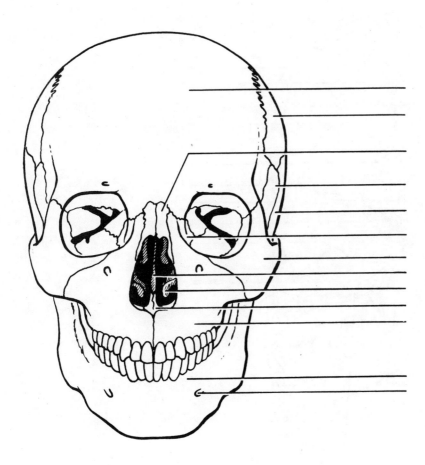

Figure 10.

D. *Sagittal Section of Skull.*

Bones

frontal
inferior concha
maxilla
occipital
palatine
parietal
temporal
vomer

Processes

crista galli
horizontal plate
internal acoustic meatus
perpendicular plate
sella turcica
sphenoid sinus
ramus

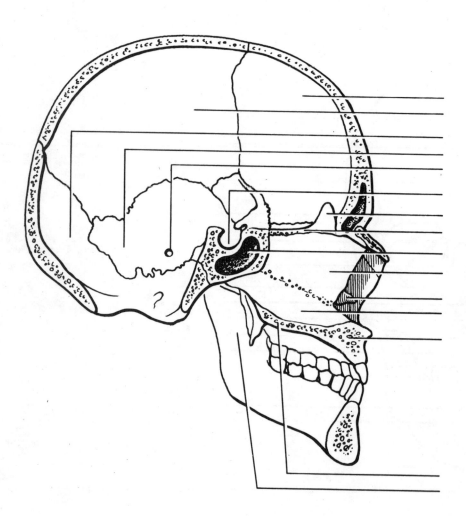

Figure 11.

E. *Base of Skull. Internal Aspect.*

Bones

frontal
occipital
parietal
temporal

Processes

crista galli
foramen magnum
great wing (sphenoid)
horizontal plate
olfactory foramen
petrous portion (temporal)
sella turcica
small wing (sphenoid)

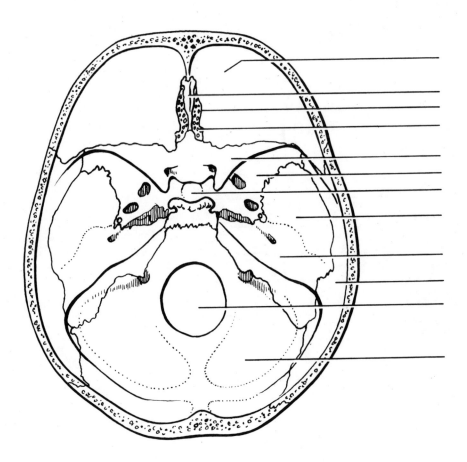

Figure 12.

F. *Base of Skull. External Aspect.*

Bones

maxillary
occipital
palatine
temporal
vomer
zygomatic

Processes

condyle
foramen magnum
mastoid
pterygoid
zygomatic arch

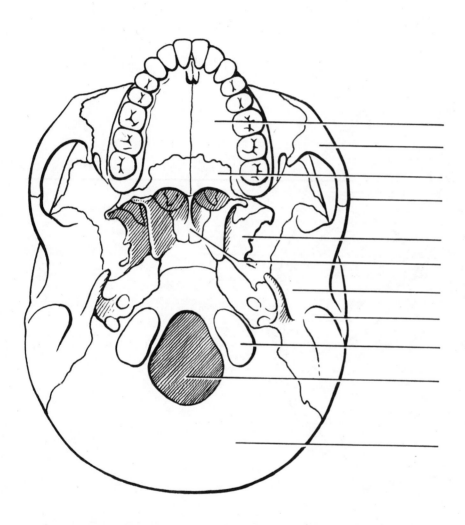

Figure 13.

Part B — **Vertebral Column and Thorax**

General Supplies:
    anatomical wall charts
    human skeletons (articulated and disarticulated)

I.   *Study of the Skeleton.*

   A.   Examine the articulated skeleton and the wall charts. Note particularly the ribs and their attachments. Look at the bony vertebral column and note the normal curves.

   B.   Using the disarticulated skeleton study the individual bones of the thorax and vertebral column. With the aid of text and reference books identify the prominent markings on these bones. Locate the bones on yourself and on your partner.

   C.   Answer the following questions:

      1.   The first cervical vertebra also is known as the ————————————————

      2.   It articulates with the ———————————————————————— bone

         above and with the second cervical vertebra or ———————————— below.

      3.   The ———————————————————————————— fills in the

         posterior portion of the pelvic girdle.

      4.   The anatomical term for "tailbone" is ————————————————

      5.   If you were given a box of assorted vertebrae and told to separate them into 3 groups what special features would help you in determining the correct classification?

         cervical ————————————————————————————

         thoracic ————————————————————————————

         lumbar ————————————————————————————

II.   *Label the following diagrams as indicated.*

A. *Cervical Vertebrae.*

body
odontoid process
spinous process

superior articular process
transverse foramen
vertebral foramen

1. *The Atlas. Superior Aspect.*

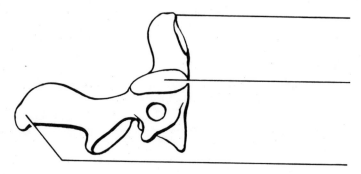

2. *The Axis. Lateral Aspect.*

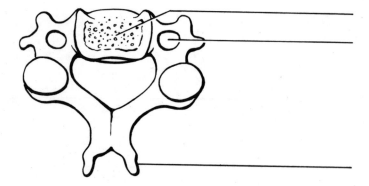

3. *Typical Cervical Vertebra. Superior Aspect.*

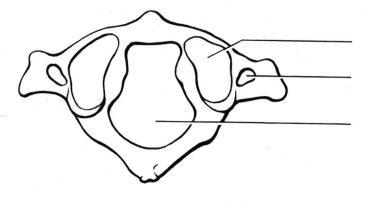

Figure 14.

B. *Thoracic and Lumbar Vertebrae.*

body	inferior articular process	spinous process
lamina	superior articular process	transverse process
pedicle	articular facets for ribs	vertebral foramen

1. *Thoracic Vertebra. Lateral Aspect.*

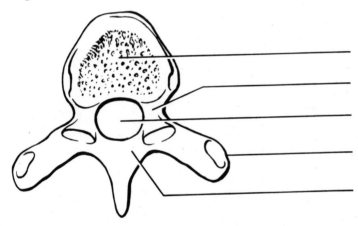

2. *Thoracic Vertebra. Superior Aspect.*

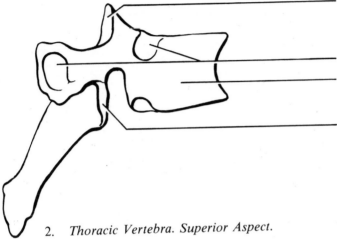

3. *Lumbar Vertebra. Lateral Aspect.*

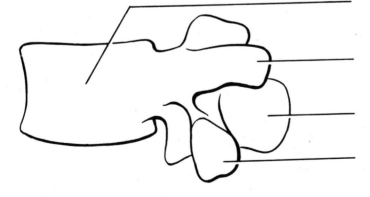

Figure 15.

C. *Sternum and Ribs. Anterior Aspect.*

With colored pencils differentiate between the vertebrosternal, vertebrochondral, and vertebral, or floating, ribs.

Label the sternum as indicated: body

manubrium

xiphoid

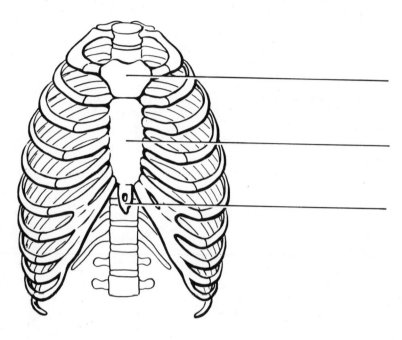

Figure 16.

III. *Application to Practical Situations.*

A. Mrs. J. R. suffered a depressed fracture at the base of her skull, and was unconscious when brought to the hospital.

1. Name the cranial bone that was injured. _____

2. What is the difference between a depressed skull fracture and a simple skull fracture?

_____

_____

3. The surgeon performed a craniotomy. In very simple terms what does this mean?

_____

B. Mr. T. H. was admitted to the hospital for correction of a deviated nasal septum.

    1.   Briefly, what was wrong with him? _____

_____

    2.   What bones form the greater part of the nasal septum? _____

_____

C. Many individuals suffer from sinusitis.

    1.   What is this condition? _____

_____

    2.   List the various air sinuses and name the cavities into which they open: _____

_____

_____

_____

    3.   What kind of membrane lines these sinuses?

_____

    4.   What are the functions of the paranasal sinuses?

_____

_____

    5.   The maxillary sinus also is known as the _____

_____

D. A young mother asks if it is dangerous to touch the "soft spots" on the head of her newborn baby. She has heard that even gentle pressure on these areas might be fatal.

    1.   How would you answer her question?

_____

_____

_____

_____

_____

2. At approximately what age do the following fontanels close:

anterior _____

posterior _____

E. Poor posture as well as disease may produce changes in the normal curvatures of the vertebral column. Identify the following conditions, and list at least one posture fault that might have been a contributing factor in each case.

kyphosis _____

_____

lordosis _____

_____

scoliosis _____

_____

F. As an aid in the diagnosis of anemia the doctor may perform a bone marrow biopsy.

1. What is this procedure? _____

2. Why is it of value as a diagnostic tool? _____

_____

G. Your neighbor's baby has a spina bifida.

1. Anatomically, what is responsible for this congenital defect? _____

_____

2. In what regions does it occur most frequently? _____

_____

LABORATORY 6

The Appendicular Skeleton

Part A — **The Extremities**

Student Supplies:
    anatomical wall charts
    human skeletons
        (articulated and disarticulated)

I. *Study of the Skeleton.*

    A. Examine the bones of the disarticulated skeleton. Compare them with the bones of the articulated skeleton.

    B. With the help of text and reference books identify the major processes on each bone.

    C. Locate these bones in your own body.
       By palpation try to identify the larger processes that lie just under the skin.

II. *Label the diagrams found on the following pages.*

    A. *PRINT* each label at the *end* of the proper line.

    B. You will find it helpful to have the individual bone at hand as you study each drawing.

42

C. *Right Clavicle and Scapula.*

acromial end
acromion process
axillary border
coracoid process
glenoid process

infraspinous fossa
sternal end
supraspinous fossa
spine
vertebral border

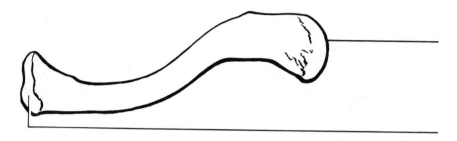

Superior Aspect of Clavicle

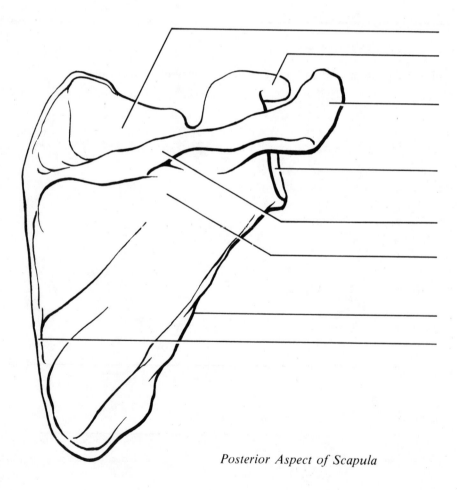

Posterior Aspect of Scapula

Figure 17.

D. *Right Humerus.*

anatomical neck   lesser tubercle
greater tubercle   medial epicondyle
head   olecranon fossa
lateral epicondyle   surgical neck

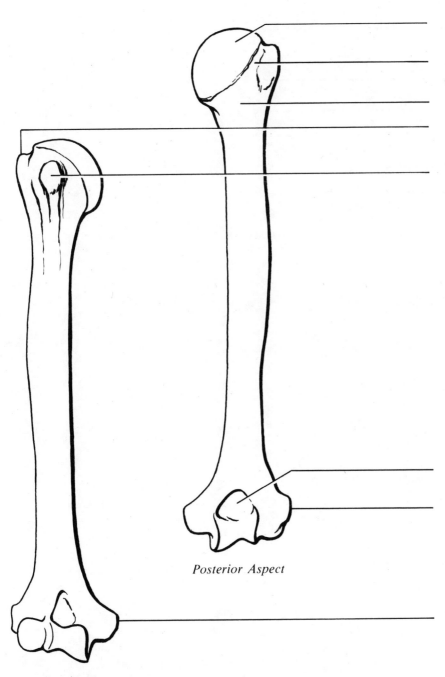

Posterior Aspect

Anterior Aspect

Figure 18.

E. *Left Radius and Ulna.*

Radius	*Ulna*
head	olecranon process
neck	semilunar notch
styloid process	styloid process
tuberosity	

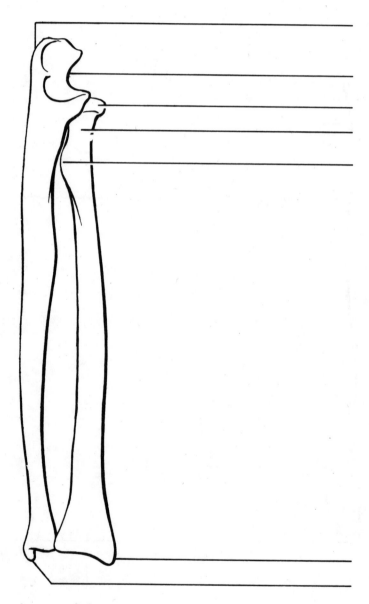

Anteromedial Aspect

Figure 19.

F.   *Posterior Aspect of Left Hand.*

carpal bones
metacarpal bones
phalanges

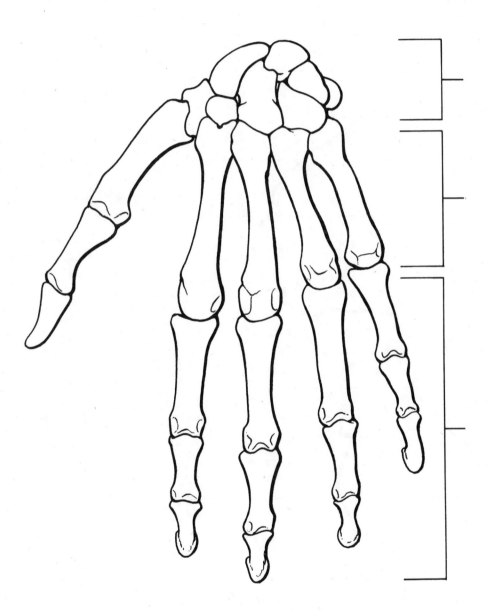

Figure 20.

46

G. *Lateral Aspect of Left Hip Bone.*

acetabulum
anterior superior iliac spine
great sciatic notch
iliac crest
ilium

ischium
obturator foramen
posterior superior iliac spine
pubis

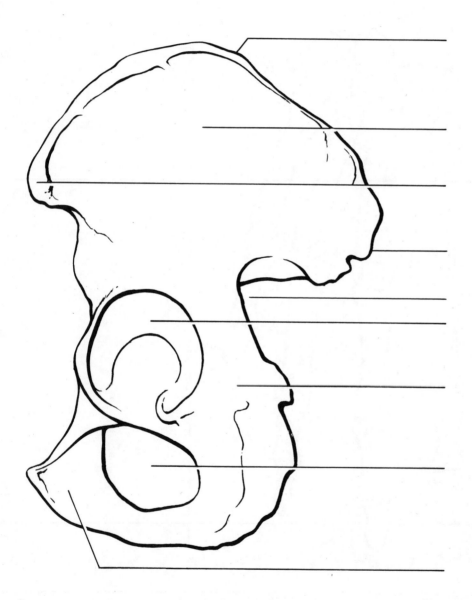

Figure 21.

H. *Left Femur.*

greater trochanter
head
intercondyloid fossa
lateral condyle
lesser trochanter

linea aspera
medial condyle
neck
patellar surface

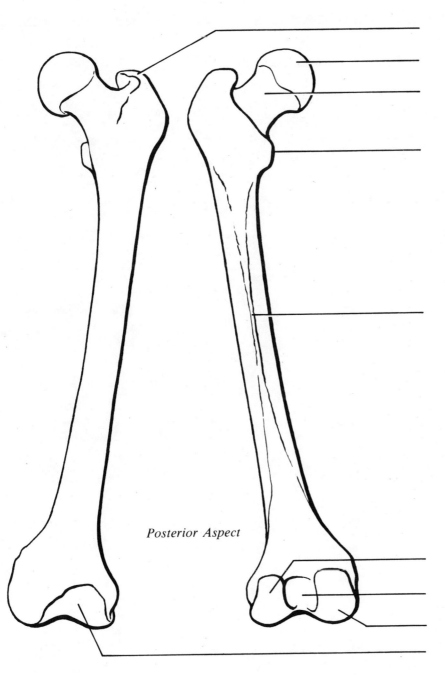

Posterior Aspect

Anterior Aspect

Figure 22.

48

I.   *Anterior Aspect of Left Tibia and Fibula.*

      intercondylar eminence
      lateral condyle
      lateral malleolus
      medial condyle
      medial malleolus
      tuberosity

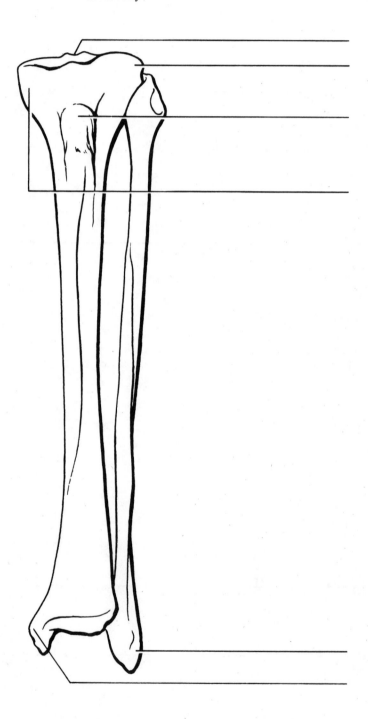

Figure 23.

J. *Superior Aspect of Left Foot.*

metatarsal bones
phalanges
tarsal bones (cuboid, navicular, cuneiforms)
    talus (astragalus)
    calcaneus

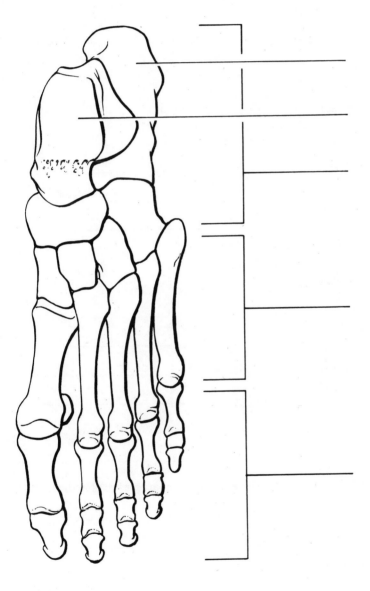

Figure 24.

50

III. *Application to Practical Situations.*

    A.   Mrs. R. B. slipped on some icy pavement. In attempting to cushion the fall she extended her right arm in such a way that she landed on the palm of her hand. This resulted in a typical *Colles' fracture.*

        1.   Describe this fracture in anatomical terms.

        _____

        2.   Give a logical explanation for the fact that a Colles' fracture most often occurs in the type of fall experienced by Mrs. R. B.

        _____

        _____

        _____

    B.   During a football game one of the players was tackled in such a way that he fell heavily on the lateral aspect of one shoulder. In spite of his pads he sustained a severe fracture.

        1.   Which bone would he be most likely to break in this type of fall? _____

        2.   Why is this bone particularly vulnerable? _____

        _____

    C.   We often say that a person has a "slipped disk."

        1.   How could this defect be more accurately described?

        _____

        _____

        2.   What are some activities that might cause this type of injury? _____

        _____

    D.   What is the common term for an excessive amount of inversion or eversion of the foot?

        _____

    E.   Why is a fracture through the shaft of the tibia more disabling than a comparable break in the fibula?

        _____

        _____

Part B — **The Articulations**

Demonstration Supplies:
    fresh beef joints (vertebral, hip, knee)
    dissecting instruments
    x-ray view box
    x-ray films of various joints (adult and child)

Student Supplies:
    articulated skeleton
    anatomical wall charts

I.   *Demonstration of Beef Joints.*

The instructor will demonstrate fresh beef joints and give you an opportunity to identify the basic structures. After you return to your seat answer the following questions:

A.   *The vertebral joints.*

    1.   What did you see between the bodies of the vertebrae? _____

    _____

    2.   What is the function of this material? _____

    _____

    3.   Name the ligament which extends over the spinous processes from C-7 to the occiput.

    _____

    4.   What ligament extends over the spinous processes from C-7 to the sacrum? _____

    5.   A series of ligaments connect the lamina, extending from sacrum to axis. What are they called?

    _____

    6.   Name the two ligaments that connect the bodies of the vertebrae. _____

    _____

    7.   Was there any movement between the bodies of the vertebrae? _____

    _____

    8.   How would you classify these joints according to the amount of motion? _____

    _____

    9.   Name two other joints in the body that are of the above type. _____

    _____

    10. What kind of movement was seen between the superior and inferior articular processes?

    _____

B.   *The hip joint.*

    1.   What bones articulate at the hip? _____

    _____

    2.   The articular surfaces are cushioned by what kind of tissue? _____

    _____

3. How does the hip joint differ from the shoulder joint even though they are both of the ball and socket type? _____

_____

_____

_____

_____

4. What bones articulate at the shoulder? _____

_____

_____

5. How would you classify these joints according to the amount of motion? _____

_____

6. These joints are lined with what type of membrane? _____

_____

7. What types of movement can be obtained at ball and socket joints? _____

_____

C. *The knee joint.*

1. What bones articulate at the knee? _____

_____

2. Explain what is meant by the term "meniscus." _____

_____

3. What structures deep within the knee joint give added strength? _____

_____

4. List the other hinge joints in our body, and name the bones that articulate in each case.

_____

_____

_____

5. How would you classify the hinge joint according to amount of motion? _____

_____

6. What types of movement are possible at hinge joints? _____

_____

7. How does the wrist joint differ from the typical hinge joint on the basis of structure and types of movement possible? _____

_____

_____

_____

8. What do we call the lubricating fluid that is found in freely movable joints? _____

_____

II. *A Study of X-ray Films.*

The instructor will demonstrate a series of x-ray films.

This will give you an opportunity to study the bones and joints as they are commonly seen in the living body.

III. *The Articulated Skeleton.*

Working in groups of four or six, examine the various joints in the articulated skeleton.

Compare these joints with those in your own body.

Review the bones and types of movement possible at each joint.

IV. *Comparison of Male and Female Pelves.*

Label the following diagrams as indicated by the lines extending from the various bones and joints. Indicate which drawing represents the female and which the male.

_____*Pelvis*___

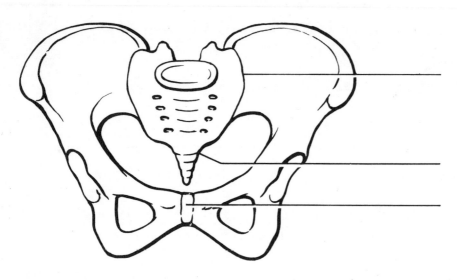

_____*Pelvis*___

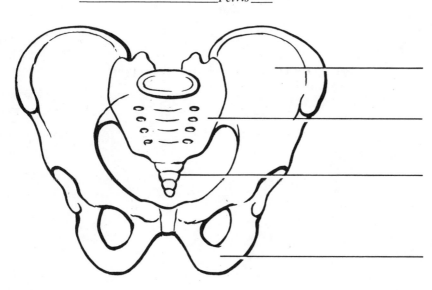

Figure 25.

Distinguish between the "true pelvis" and the "false pelvis."

_____

_____

_____

LABORATORY 7

Introduction to the Muscular System

Part A — **Comparative Physiology**

Demonstration Supplies:
   live frog
   dissecting sct
   ring stand and clamp
   2 small beakers
   Ringer's solution
   distilled water

Student Supplies:
   live frog for each two students
   dissecting instruments and trays
   sodium chloride crystals
   Ringer's solution
   small beakers
   litmus paper

I. *A Study of Living Frog Muscle.*

   A. *Preparation of the frog.*

      1. The instructor will pith or anesthetize each frog.

      2. Place the frog ventral side up in your dissecting tray.

      3. Make the usual mid-line incision through the skin, muscular body wall, and the sternum as you did in Laboratory 1, page 1.

      4. Pin the frog into the wax in your tray so that the thoracic and abdominal viscera are well exposed.

   B. *Study of cardiac and visceral muscle.*

      1. Examine the beating heart of your frog. The serous sac which surrounds the heart is called

      _____

      2. Using forceps, pick up this sac and snip it open with your scissors.

      3. Grasp the apex (pointed end) of the heart with your forceps and gently pull it free of the sac. Note the large blood vessels at the base of the heart.

      4. With scissors cut through the blood vessels and remove the entire heart.

      5. Put the heart in a small beaker of Ringer's solution and examine it at intervals.

         a. Does it continue to beat? _____

         b. What special property of muscle tissue does this illustrate? _____

6. With forceps pinch the upper end of the stomach. Watch closely for several seconds.

    a. What did you see? _____

    b. What is this called? _____

    c. It was due to the contraction of _____
       muscle tissue.

C. *Study of skeletal muscle tissue.*

1. With heavy scissors cut the frog into cranial and caudal portions. Make this transverse section just above the pubis.

2. Separate the hind legs by cutting longitudinally through the pubis and spinal column.

3. Remove the skin from both hind legs.

4. Remove the anterior thigh muscle from one of the legs. Cut it and touch the cut ends to litmus paper.

    a. What was its reaction? _____

    b. Explain: _____

    _____

5. On the dorsal aspect of the other hind leg, locate the longitudinal groove between the posterior thigh muscles.

6. Using a blunt probe carefully separate the muscles along the groove until you find the sciatic nerve.

7. Slip your probe under this nerve and leave it there.

8. With forceps pinch the nerve near its proximal end while watching the gastrocnemius muscle.

    a. What happens? _____

    b. Could you see any change in the nerve? _____

9. Now pinch the body of the gastrocnemius muscle instead of the nerve.

    a. What happens? _____

    b. How do you account for the difference in response?

    _____

    _____

    _____

10. Place a few crystals of sodium chloride on the thigh muscles.

    a. What happens? _____

    b. Why? _____

D. Wrap the frog in old newspaper and discard in appropriate waste container.

E. Wash and dry all equipment before returning it to the supply table.

II. *Demonstration Frog.*

   A. The instructor has cut the spinal nerves leading to muscles in one of the hind legs of a frog. Examine this frog as it hangs suspended from the ring stand.

     1. Which hind leg has no nerve supply? _____

     2. How can you determine this? _____

        _____

     3. What happens when you pinch the toes of each foot? Why?

        _____

        _____

   B. Examine the strips of muscle tissue which have been soaking in distilled water and in Ringer's solution.

     1. Was there any change in the tissue soaked in Ringer's solution? _____

     2. Explain: _____

        _____

     3. What happened to the tissue soaked in distilled water?

        _____

     4. Explain: _____

        _____

Part B — **Human Physiology**

Demonstration Supplies:
   microscopes
   prepared slides of muscle tissue

Student Supplies:
   tape measures
   blood pressure cuffs

I. *Microscopic Study of Human Muscle Tissue.*

   A. Examine the prepared slide of skeletal muscle tissue.

     1. Locate: sarcolemma, sarcoplasm, striae, and nuclei.

     2. Sketch and label a few cells.

3. What is the function of this tissue? _____

_____

B. Examine the prepared slide of visceral muscle tissue.

   1. Locate: cell membrane and nucleus

   2. Sketch and label a few cells.

   3. Where would this tissue be found in your own body?

      _____

   4. What is its function? _____

      _____

C. Examine the prepared slide of cardiac muscle tissue.

   1. Locate: cell membrane, nucleus, and intercalated disks.

   2. Sketch and label a few cells.

   3. What is the significance of the latticework arrangement of cardiac muscle cells? _____

      _____

D. Examine the prepared slide of motor end-plates.

   1. Locate: skeletal muscle fibers, nerve fibers, and end-plates.

   2. What is a motor end-plate or neuromuscular junction? _____

      _____

   3. What is the function of acetylcholine? _____

      _____

   4. What is the function of cholinesterase? _____

      _____

II. *Macroscopic Studies of Muscle Contractions.*

A. Have your partner measure the circumference of your arm with the forearm extended and with it flexed. The tape measure should surround the body of the biceps brachii; mark the overlying skin so that the tape measure is applied at the same place each time.

   1. Record the measurements:

      a. with forearm extended: _____ inches.

      b. with forearm flexed: _____ inches.

2. What is responsible for the change in size?

_____

B. Hold a book in your hand while *rapidly* flexing the forearm at least 50 times.

1. Measure the circumference of your arm once again.

2. What has happened? _____

3. Explain: _____

_____

C. Partially flex your left forearm while resting the elbow on your desk. Support the left hand at the wrist with the fingers of your right hand.

1. Vigorously flex and extend the fingers of your left hand until fatigued.

How long did it take? _____ minutes.

2. Rest your hand until it is recovered.

3. Now have your partner apply a blood pressure cuff just above the left elbow. The cuff should be inflated until the radial pulse can no longer be palpated.

4. Repeat the flexion and extension exercise of the fingers until again fatigued.

How long did it take? _____ minutes.

5. In which exercise was fatigue experienced most rapidly?

_____

6. Explain: _____

_____

III. *Application to Practical Situations.*

A. Pathologic Contractions.

1. What is meant by the general term "muscle spasm"? _____

_____

2. Distinguish between tonic and clonic spasms. _____

_____

_____

3. What is fibrillation? _____

_____

B. Mr. Johns has *flaccid paralysis* of his lower extremities.

What does this mean? _____

_____

C. A youngster who lives in your neighborhood has *spastic paralysis*.

Explain: _____

_____

D. When it is cold weather we often experience the sensation of shivering.

Explain this on a physiological basis: _____

_____

_____

E. What is muscle tone? _____

_____

1. Maintenance of muscle tone depends upon _____

_____

2. When is muscle tone lowest? _____

_____

F. In the operating room, patients are often given drugs that are curare-like compounds.

1. What is the function of these drugs? _____

_____

2. Upon what specific structures do they act?

_____

G. While working at night you walk from a lighted hall into a dark room. Just as you do so, there is a loud crash of breaking glass.

Which of the following responses would occur first?

1. The pupils of your eyes will dilate._____

2. You will jump with alarm._____

Explain your choice: _____

_____

H. Distinguish between muscular atrophy and muscular hypertrophy.

_____

_____

Introduction to the Skeletal Muscles.

Label the following diagrams as indicated by the various label lines.

Anterior View

adductor group
biceps brachii
external oblique
pectoralis major
quadriceps femoris
rectus abdominis
sartorius
serratus anterior
sternocleidomastoid
tibialis anterior

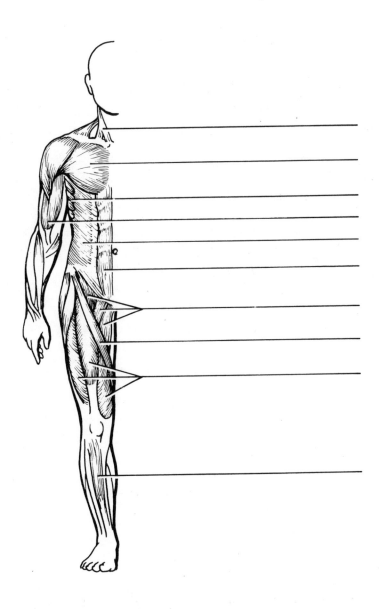

Figure 26.

62

Posterior View

deltoid
gastrocnemius
gluteus maximus
hamstrings
latissimus dorsi
tendon of Achilles
teres major
trapezius
triceps

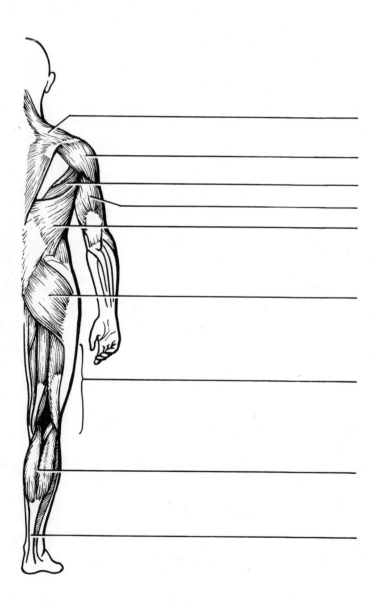

Figure 27.

LABORATORY 8

The Major Skeletal Muscles

Demonstration Supplies:
    embalmed cat (muscles dissected)

Student Supplies:
    anatomical wall charts
    reference books

I. *Demonstration of Cat.*

The instructor will point out the major skeletal muscles as they are found in an embalmed cat.

II. *Muscles of the Head and Neck.*

A. *Facial expression.*

1. What muscle wrinkles the forehead when you look at your partner in surprise? _____

2. Close your eyes tightly and then open them. What muscles did you use?

a. to open the eyes _____

b. to close the eyes _____

3. With your jaws closed, what muscles would you use

a. to bare your teeth? _____

b. to purse your lips? _____

c. to pull the corners of your lips down? _____

_____

B. *The eyeball.*

1. What muscles are used to

a. look at the tip of your nose? _____

b. glance at the floor? _____

c. look at the ceiling? _____

2. When reading a book what muscles move the eyeball

a. from right to left? _____

b. from left to right? _____

C. *Mastication.*

    1. Pretend that you are chewing some food. What muscles did you contract to

        a. produce side-to-side motion? _____

        b. close the jaws? _____

    2. What muscle helps us to keep food from collecting between the teeth and cheeks? _____

        _____

    3. Good manners dictate that the lips remain closed while chewing food. How can we do this?

        _____

D. *The tongue.*

    1. What are the functions of the intrinsic muscles? _____

        _____

    2. When protruding the tongue you must contract the _____

        _____

    3. To retract the tongue contract the _____

E. *Movement of the head.*

    1. Tilt the head toward one shoulder while holding your hand on the muscles of your neck.

    a. What muscles produced this lateral flexion? _____

        _____

    b. What muscles enable you to bring the head back to original position? _____

        _____

    2. Move your head down toward your chest.

        a. What is this movement called? _____

        b. Muscles used: _____

        c. The antagonistic muscles are _____

        _____

    The motion produced is called _____ of the head.

III. *Muscles of the Chest.*

    A. What muscles are used in regular quiet breathing:

        1. to inhale? _____

        2. to exhale? _____

    B. When suffering with a heavy cough it is necessary to inhale deeply and then expel air forcibly. What *additional* muscles do we now bring into action to:

        1. inhale deeply? _____

                       _____

        2. exhale forcibly? _____

                       _____

IV. *The Anterolateral Abdominal Wall.*

    A. Relax your abdominal muscles completely.

        1. How would you describe the general outline of the abdomen? _____

                       _____

        2. Does this contribute to good posture? _____

    B. Contract the abdominal muscles and note the change in contour.

        Describe the general outline: _____

        _____

    C. People who must wear girdles or corsets may develop weak abdominal wall muscles.

        1. Why? _____

        2. How could such weakness be prevented even though the artificial support must be worn?

        _____

    D. What is the general action of the abdominal wall muscles in vomiting, defecation, etc.? _____

    _____

V. *Muscles of the Back.*

    A. Good back rubs improve circulation to the musculature as well as to the skin of bedridden persons.

        1. List the superficial muscles that are massaged during a complete back rub: _____

        _____

        _____

2. What is the deep muscle that lies just lateral to the vertebral column? _____

3. What is the action of this muscle? _____

_____

B. Describe briefly good body mechanics in

1. lifting heavy objects from the floor: _____

_____

2. pulling a mattress up toward the head of the bed: _____

_____

VI. *Muscles of the Extremities.*

A. Move your upper extremity in every possible direction. Opposite the following positions record the muscles that are prime movers:

1. arm flexed and adducted _____

2. arm extended and adducted _____

3. arm abducted _____

4. forearm flexed _____

5. forearm extended _____

6. palm down _____

7. palm up _____

8. hand clenched _____

9. hand open _____

B. Follow the same procedure with your lower extremity:

1. thigh flexed _____

2. thigh extended _____

3. thigh abducted _____

4. thigh adducted _____

5. leg flexed _____

6. leg extended _____

7. foot flexed (dorsiflexion) _____

8. foot extended (plantar flexion) _____

VII. *Application to Practical Situations.*

    A. Mr. S. R. has been admitted to the hospital for treatment of a *wry neck.*

        1. What is the medical term for this defect? _____

        2. The _____ muscle frequently is at fault.

        3. Briefly describe Mr. S. R.'s appearance: _____

        _____

    B. Mrs. B. N. submitted to a *radical mastectomy* for carcinoma of the right breast. Part of this operation involves the removal of pectoral muscles as well as the breast tissue.

        1. What are the functions of the pectoralis major?

        _____

        2. Of the pectoralis minor? _____

        3. List at least 2 other muscles that can take over for the excised pectoral group. Give the action of each.

        _____

        _____

        4. Why is the combing of her own hair an excellent postoperative exercise for Mrs. B. N.?

        _____

    C. During an appendectomy an incision is made through the abdominal wall in the right iliac or inguinal region. List in sequence the layers of the wall that will be encountered by the surgeon before he reaches the appendix.

        _____

        _____

        _____

    D. Miss P. C., a ballet dancer, was struck by an automobile. Her left tendon of Achilles was completely severed. Explain the full significance of this injury: _____

        _____

        _____

    E. Mr. Y. N. states that he has an *indirect inguinal hernia.*

        1. Anatomically, how would you describe this defect?

        _____

        2. What is the basic etiological factor in all hernias?

        _____

3. What are some contributing factors? _____

_____

4. List the weak places in the abdominal wall:

_____

VIII. *Complete the Following Review Chart:*

Muscle	Main Action	Major Antagonist
biceps brachii		
deltoid		
diaphragm		
gastrocnemius		
gluteus maximus		
hamstrings		
inferior rectus		
lateral rectus		
latissimus dorsi		
orbicularis oculi		
peroneus longus		
sacrospinalis		
sternocleidomastoid		

LABORATORY 9

Introduction to the Nervous System

Part A — **Human Nerve Tissue**

Demonstration Supplies:
    prepared slides of nerve tissue
    microscopes
    2 x 2 projector slides (optional)

Student Supplies:
    reference books

I. *A Study of Microscopic Structures.*

    A. Examine the prepared slide of myelinated nerve fibers.

        1. Locate: axis cylinder, myelin sheath, nodes of Ranvier.

        2. Sketch and label one fiber:

        3. What is the function of the myelin sheath? _____

            _____

        4. What is the common name for myelinated fiber groups in the brain and spinal cord?

            _____

        5. What do we call unmyelinated fiber groups in the brain and spinal cord? _____

            _____

        6. Which type of fibers predominate in our cranial and spinal nerves? _____

            _____

    B. Examine the prepared slide of neurons as seen within the spinal cord.

        1. Locate: cell bodies, nuclei, and cell processes.

        2. Sketch and label a few cells:

        3. In what part of the spinal cord are the cell bodies to be found? _____

        4. Functionally speaking, what types of neurons have their cell bodies within the cord?

            _____

C. Examine the prepared slide of neurons as found in the cerebrum.

    1. Locate: cell bodies and cell processes.

    2. Sketch and label as few cells:

    3. What special term is used in reference to the surface layer of gray matter of the cerebrum?

    _____

D. Examine the prepared slide of a spinal ganglion.

    1. Locate: cell bodies and cell processes.

    2. Specifically, where are these ganglia located?

    _____

    3. Functionally speaking, what type of neuron has its cell body in this area? _____

    4. What other types of ganglia are found within the body?

    _____

E. Examine the prepared slide of a nerve in cross-section.

    1. Locate: nerve fibers, endoneurium, perineurim, epineurium, and small blood vessels.

    2. Sketch and label this unit:

    3. What is a motor nerve? _____

    _____

    4. What is a sensory nerve? _____

    _____

    5. What is a mixed nerve? _____

    _____

F. Review the prepared slide of a motor end-plate as studied in Laboratory 7.

    1. What is another name for this structure? _____

    _____

II. *Label the following diagram: Motor Neuron*

axon   Nissl bodies   neurolemma sheath
dendrites   nucleus   node of Ranvier
neurofibrils   myelin sheath   motor end-plate

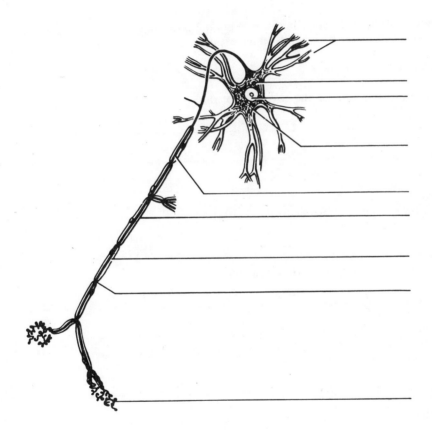

Figure 28.

Part B — **The Spinal Cord and Nerves**

Demonstration Supplies:
> embalmed cat (cord and nerves dissected)
> anatomical wall charts
> strychnine sulfate solution (0.5%)
> 5 cc. syringe and hypodermic needle

Student Supplies:
> live frog for each 4 students
> dissecting trays
> ring stands and clamps
> dilute HCl (2.5 to 3.0%)

> cotton tipped applicators
> beakers (400 ml)
> thumb forceps (toothed)

I. *Anatomy of the Spinal Cord and Spinal Nerves.*

  A. Demonstration of embalmed cat.

  The instructor will meet with small groups to point out the anatomical structures in a cat which has the spinal cord and major peripheral nerves exposed.

  B. Answer the following questions:

  1. Briefly locate and describe the coverings of the cord:

  _____

  _____

  _____

  2. What term is used in reference to these coverings?

  _____

  3. The spinal cord is wider in certain regions.

  a. Where? _____

  b. Why? _____

  4. What is the cauda equina? _____

  _____

  5. The spinal cord extends from the _____ downward

  to the level of _____

  6. The dura mater extends downward to the level of the _____

  7. Spinal nerves arise from the cord by two roots.

  a. Which contains afferent fibers? _____

  b. Which contains efferent fibers? _____

8. How do spinal nerves leave the bony vertebral column?

    _____

9. After leaving the vertebral column, each spinal nerve gives off 2 primary branches, the anterior and posterior rami. What is the general destination of each?

    _____

    _____

    _____

10. List the four largest nerves arising from the brachial plexus and briefly give their distribution:

    _____

    _____

    _____

11. From what plexus do the following nerves arise and what is their general distribution?

    a. sciatic: _____

    b. femoral: _____

    c. obturator: _____

II. *A Study of Reflex Action.*

A. The instructor will pith the brain of your frog, but the spinal cord will be left intact.
   (Note: occasionally the frog will not respond for several minutes following the destruction of its brain. This is due to shock, but the frog feels no pain and spinal reflex activity will soon be resumed.)

    1. Place the frog dorsal side up on the moistened surface of your dissecting tray.

    2. Straighten its hind legs.

        a. Does it flex them again? _____

        b. What kind of reflex is this? _____

    3. Suspend the frog from a ring stand by placing its lower jaw in a utility clamp. FASTEN SECURELY!

        a. Pinch the toes of one foot.

            Result? _____

        b. With applicator, apply dilute acid to one foot.

            Result? _____

        c. Without taking frog down from ring stand, remove acid by rinsing foot in a beaker of water.

        d. Apply acid to dorsal body wall.

            Result? _____

    e.   Remove acid with beaker of water as before.

    f.   Apply acid to the ventral body wall.

        Result? _____

    g.   Remove acid as before.

4.   What general type of reflex has been demonstrated in the foregoing exercises?

_____

5.   The instructor will inject your frog with 0.4 cc. of a 0.5% strychnine sulfate solution. This drug may be injected into the peritoneal cavity or under the skin of the throat.

    a.   Wait 4 or 5 minutes and then repeat the previous types of stimulation.

        Result? _____

    b.   Strike the ring stand firmly with your forceps.

        Result? _____

6.   What general type of reflex has just been demonstrated?

_____

7.   What is the action of strychnine in this case?

_____

_____

B. Normal human reflexes.

    1.   Standing near the window (or using a flashlight) test the response of your partner's eyes to light. Cover the eyes with a card for a few seconds before starting this exercise.

        a.   What is the *first* response to sudden bright light?

_____

        b.   The second? _____

        c.   Shade one eye but allow the other to remain uncovered. What do you notice in regard to pupils?

_____

        d.   What are these reflexes called? _____

_____

    2.   Stand in front of your partner and watch the pupils of the eyes very closely while *firmly* pinching the nape of the neck (ciliospinal reflex).

        Result? _____

3. Have partner sit down with the right leg comfortably crossed over the left. Strike the right patellar tendon with the side of your hand.

   a. What happens? _____

   b. What is this called? _____

4. While your partner kneels on a chair strike one of the Achilles' tendons with the side of your hand.

   a. What happens? _____

   b. What is this called? _____

5. List some examples of protective reflexes: _____

   _____

III. *Diagram of the Spinal Cord in Cross Section. (on following page)*

   A. Identify the structures indicated by label lines:

anterior horn	lateral horn
central canal	posterior horn
dorsal median septum	spinal nerve
dorsal root	ventral median fissure
dorsal root ganglion	ventral root

   B. Record the general function of each numbered tract:

Name of Tract	General Function
1. f. gracilis ⎤ posterior	_____
2. f. cuneatus ⎦ columns	_____
3. dorsal spinocerebellar ⎤	_____
4. ventral spinocerebellar ⎦	_____
5. lateral spinothalamic	_____
	_____
6. ventral spinothalamic	_____
	_____

7. lateral corticospinal
   (crossed pyramidal)

   _____

8. ventral corticospinal
   (uncrossed pyramidal)

   _____

9. extrapyramidal
   (rubrospinal)

   _____

C. Diagram a three-neuron reflex arc.

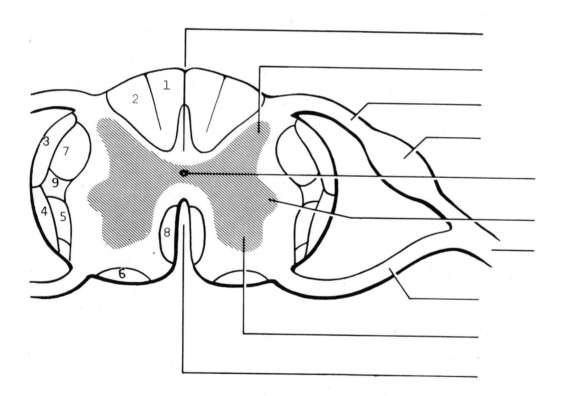

Figure 29.

IV. *Application to Practical Situations.*

   A. A doctor is going to do a lumbar puncture on one of his patients to secure a sample of cerebrospinal fluid.

   1. The patient must lie on one side, head and trunk flexed, knees pulled up toward his chin. WHY?

   _____

   _____

   _____

   2. Between which two vertebrae will the doctor insert the needle. WHY?_____

   _____

   3. From what space will he secure the sample of fluid? _____

   _____

   4. As the fluid is withdrawn you note that it has a marked red color. Is this normal? Explain:

   _____

   _____

   B. Personnel working in the emergency room receive two patients who were in an automobile accident. One is dead on arrival at the hospital, having suffered a transection of the spinal cord at the level of C-2. The other patient suffered a similar injury but at the level of C-6, and he is still alive.

   1. Explain briefly, in terms of phrenic nerve origin and function, why one injury was fatal while the other was not: _____

   _____

   _____

   _____

   2. What general symptoms would you expect the second patient to present? _____

   _____

   _____

   3. What are his chances for recovery? _____

   4. What factors are essential in the regeneration of nerve fibers? _____

   _____

   _____

C. What is meant by "extradural anesthesia?" _____

_____

D. Why does an amputee sometimes complain of pain or itching in the missing extremity?

_____

_____

_____

E. Injury to what nerves will product the following symptoms?

   1. wrist drop: _____

   2. foot drop: _____

   3. inability to extend leg at knee: _____

   4. inability to flex leg at knee: _____

F. A positive Babinski reflex may indicate injury to which spinal cord tracts? _____

_____

G. Why are you unaware of the presence of your wristwatch after you have worn it for some time?

_____

_____

H. Mrs. B. B. was suffering intractable pain as a result of carcinoma. To relieve her pain the neurosurgeon performed a chordotomy.

   1. Briefly what did this operation involve? _____

_____

   2. Why might it be expected to give relief from pain? _____

_____

   3. What are the physical implications in regard to the use of hot water bottles or heating pads?

_____

I. Why should injections be given into the lateral aspect of the deltoid region rather than into the posterior aspect?

_____

LABORATORY 10

The Brain and Cranial Nerves

Part A — **The Brain**

Demonstration Supplies:
fresh sheep brains (2 per lab section)
preserved human brain (optional)
anatomical wall charts

Student Supplies:
preserved sheep brain for each two students
dissecting instruments and trays
model brains (regular plus torso model)
reference books

I.   *Study of Preserved Sheep Brains.*

   A. *Preliminary examination.*

      1.   Identify the dura mater, falx cerebri, and tentorium cerebelli while viewing the brain from above.

      2.   Examine the base of the brain and identify the following:

           olfactory bulbs   pituitary gland
           optic chiasma   trigeminal nerves

      3.   To facilitate further study, remove the dura mater by cutting through the tentorium cerebelli and then gently pulling the entire membrane free of the brain. The pituitary gland may also be removed.

      4.   With the help of text and reference books continue to identify surface features as follows:

           cerebrum and cerebellum
           midbrain, pons, medulla
           convolutions or gyri
           the major fissures:
               longitudinal and transverse
               lateral (Sylvius) and central (Rolando)
           lobes of cerebrum:
               frontal and parietal
               temporal and occipital
           cranial nerve origins (when possible)

   B. *Dissection of the brain.*

      1.   Gently separate the parietal lobes along the longitudinal fissure until you can see the band of white fibers connecting the cerebral hemispheres.

      2.   Using a sharp scalpel carefully make a mid-line incision through the remainder of the cerebrum, brain stem, cerebellum, and spinal cord. When finished you will have two separate halves of brain and cord.

3. Looking at the interior of the brain identify the following:

> corpus callosum
> lateral ventricles
> choroid plexus
> interventricular foramen (of Monro)
> third ventricle
> thalamus
> pineal body
> cerebral aqueduct (of Sylvius)
> fourth ventricle
> cerebellum and arbor vitae
> midbrain, pons, and medulla

4. Make a coronal section of one cerebral hemisphere in the region of the thalamus. Identify:

> cortex
> corpus callosum
> lateral ventricle
> thalamus
> projection fibers

C. Clear your desk.

1. Wrap remnants of brain tissue in newspaper.
    Discard in appropriate waste container.

2. Wash and dry equipment before returning it to the supply table.

II. *Study of Fresh Sheep Brains.*

The instructor will section two fresh specimens and place them on the demonstration table. Examine them carefully and compare with the preserved brains in regard to texture.

III. *Demonstration of Human Brain.* (optional)

The instructor will meet with small groups and point out various anatomical features related to the preserved human brain. You will be expected to answer questions during this demonstration.

IV. With the help of reference books, wall charts, and brain models answer the following questions:

A. What is the correct anatomical term used in reference to:

1. ring of arteries at base of brain? _____

2. fold of dura in longitudinal fissure? _____

3. endocrine gland at base of brain? _____

4. bridge of white fibers connecting the cerebral hemispheres? _____

_____

5. network of blood vessels giving rise to cerebrospinal fluid? _____

_____

6. mass of gray matter in the walls of the third ventricle? _____

_____

7. center for sleep and temperature regulation? _____

8. center for control of motor speech? _____

B. What did you notice regarding the texture of fresh brain tissue? _____

_____

C. Why is this texture of significance to the brain surgeon? _____

_____

D. Trace the pathway of cerebrospinal fluid from the lateral ventricles to the subarachnoid space of the spinal cord:

lateral ventricles

↓

E. What are the functions of cerebrospinal fluid? _____

_____

V. *Diagrams of the Human Brain.*

A. *Lateral Aspect.*

1. Print at the end of the appropriate lines:

central fissure (Rolando)   occipital lobe
cerebellum   parietal lobe
frontal lobe   temporal lobe
lateral fissure (Sylvius)   transverse fissure

2. Identify the various areas of functional localization by shading with colored pencils.

motor area	—	dark green
premotor area	—	light green
prefrontal cortex	—	brown
general sensory area	—	red
Broca's area	—	yellow
visual area	—	orange
auditory area	—	blue

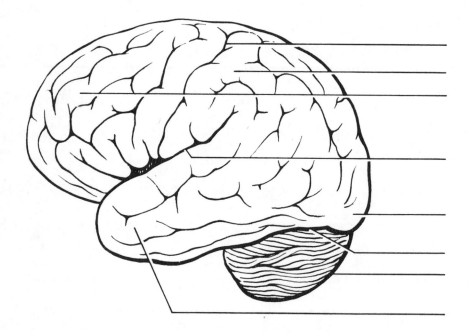

Figure 30.

B. *Sagittal Section of Brain.*

arachnoid villus
cerebral aqueduct
    (of Sylvius)
central canal of cord
cerebellum
corpus callosum
dura mater
foramen of Magendie
fourth ventricle
interventricular foramen
    (of Monro)

lateral ventricle
medulla oblongata
midbrain
pia mater
pons
spinal cord
subarachnoid space
subdural space
superior sagittal sinus
third ventricle

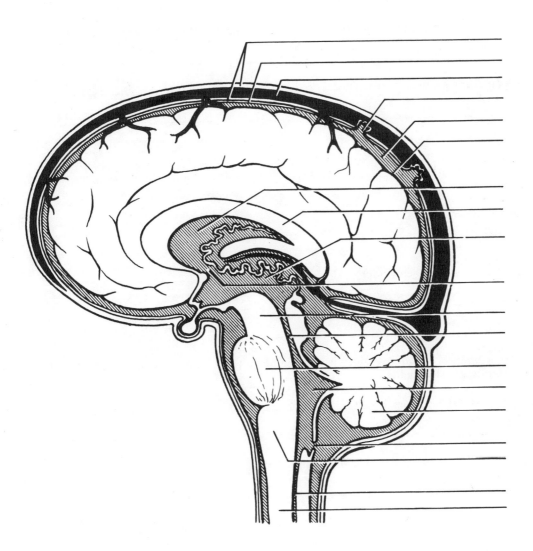

Figure 31.

VI. *Application to Practical Situations.*

    A. Why is depressed skull fracture at the base of the brain often more dangerous than one in the frontal region?

           _____

    B. Upon completion of a physical examination Dr. B.S. writes the following diagnosis on the patient's chart:

           cerebral hemorrhage.   *right hemiplegia.*

        1. How would you explain this diagnosis in simple terms?

               _____

               _____

        2. The paralysis most probably is due to injury of projection fibers in what specific area?

               _____

        3. Which side of the brain is involved? _____

            Explain: _____

               _____

    C. Jimmy has *hydrocephalus.* The doctor plans to do a ventriculoatrial shunt with polyethylene tubing and a one-way valve.

        1. What is hydrocephalus? _____

               _____

        2. Briefly, what does the doctor hope to do for Jimmy? _____

               _____

               _____

    D. Mr. D. A. has a subdural hematoma which is exerting pressure on the area immediately anterior to the right central fissure (Rolando).

        1. What is a subdural hematoma? _____

               _____

        2. The lesion would produce what general symptom? _____

               _____

        3. Which side of the body would be involved? _____

Part B — **The Cranial Nerves**

I. *A Study of Cranial Nerve Distribution and Function.*

    A. Using brain models and reference materials note the points at which the various cranial nerves arise from the base of the brain.

    B. What is the significance of the Roman numerals assigned to the cranial nerves?

    _____

    C. Record the names of the cranial nerves which are involved in the following activities and sensations:

        1. Elevate your shoulders _____

        2. Read a book _____

        3. Smell perfume _____

        4. Toothache in upper jaw _____

        5. Smile _____

        6. Listen to the radio _____

        7. Stick your tongue out _____

        8. Chew gum _____

        9. Slowing of heart beat _____

        10. Sensation of thirst _____

        11. Cinder in your eye _____

        12. Motion sickness _____

II. *Application to Practical Situations.*

    A. Mrs. L. Y. has a brain tumor. Her symptoms include: ptosis (drooping) of the upper eyelid, dilated pupil, eyeball turned downward and outward. Which nerve is involved?

    _____

    B. What is tic douloureux? _____

    _____

    C. Exposure to cold may cause a temporary paralysis of the facial nerve. What *major* symptoms would be presented in such a case?

    _____

    _____

    D. Mr. X. Y. gives a history of drinking wood alcohol instead of grain alcohol. What will be his chief complaint?

    _____

III. *Review Chart for Study of the Cranial Nerves:* (optional)

Cranial Nerve	General Distribution and Function
I.	
II.	
III.	
IV.	
V.	
VI.	
VII.	
VIII.	
IX.	
X.	
XI.	
XII.	

LABORATORY 11

The Autonomic System

Demonstration Supplies:
  embalmed cat (autonomics dissected)
  2x2 slides of entire nervous system (review)
  projector

Student Supplies:
  text and reference books
  human torso model
  anatomical wall charts

I. *Autonomics in the Cat.*

The instructor will meet with small groups and demonstrate autonomic nerves and ganglia in the embalmed cat.

II. *Autonomics in the Human.*

  A. Study the wall charts and torso model as a means of visualizing the autonomic structures of the human body.

  B. Answer the following questions:

  1. What is the general function of the parasympathetic system? _____

  _____

  2. What is another name for this system?

  _____

  3. Which cranial nerves carry parasympathetic fibers and what is their general distribution?

  _____

  _____

  _____

  _____

  _____

  4. Which sacral nerves carry parasympathetic fibers and what is their general distribution?

  _____

  _____

5. Where are parasympathetic ganglia located and what are they called? _____

_____

6. What is the general function of the sympathetic system?

_____

_____

7. What is another name for this system? _____

8. Where do the preganglionic fibers of the sympathetic system have their origins? _____

_____

9. There are two general types of sympathetic ganglia. Name them and give their location.

_____

_____

10. What is the function of the great autonomic plexuses?

_____

_____

11. What do we call the fibers which extend from any autonomic ganglion to one of the organs of the body? _____

12. The splanchnic nerves are composed of preganglionic fibers.

    a.   They are part of the _____ system of autonomics.

    b.   These fibers have synaptic connections with postganglionic fibers in the _____

_____

13. What is norepinephrine (sympathin) and where is it formed?

_____

_____

14. What is acetylcholine and where is it formed? _____

_____

_____

III. *Application to Practical Situations.*

    A. One treatment for gastric ulcer is an operative procedure called vagotomy. What effect may this have?

    _____

    _____

    B. Mr. M. P. has peripheral vascular disease. Owing to the reduced blood supply, he has a persistent ulcer on his right leg. His doctor plans to perform a lumbar sympathectomy.

        1. Briefly, what is lumbar sympathectomy? _____

        _____

        2. How will it benefit this patient? _____

        _____

    C. Miss D. J. has undergone sympathetic ganglionectomy to relieve her hyperhydrosis.

        1. What is hyperhydrosis? _____

        2. Why may ganglionectomy provide relief? _____

        _____

    D. Why is epinephrine of value in the following cases?

        1. Asthmatic attacks _____

        2. Cardiac collapse _____

        3. Combined with Novocain for minor surgery _____

        _____

    E. Why does it take just a few seconds for the body to mobilize its reserves in case of an emergency but many minutes to "calm down" when the crisis is over? _____

    _____

    _____

IV. *Autonomic Pathways.* (optional)

Using colored pencils, show the general placement of autonomic neurons and synapses related to the stomach and to visceral muscle in the skin.

preganglionic fibers
    parasympathetic —   brown
    sympathetic   —   red

postganglionic fibers
    parasympathetic —   blue
    sympathetic   —   green

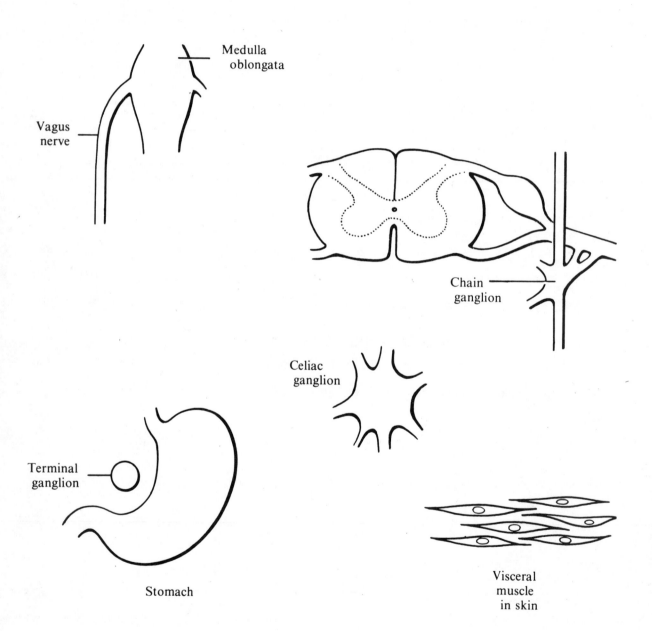

Figure 32.

LABORATORY 12

The Special Senses

Part A — The Eye

Demonstration Supplies:
    3 beakers and tongue blades
    mineral oil, water
    human skulls

    model of eye
    anatomical wall charts

Student Supplies:
    fresh beef eye for each two students
    dissecting instruments and trays
    watch glasses
    test tubes

I.   *Structure of the Human Eye.*

    A. Observe your partner's eye. With the aid of charts and text identify the following:

        conjunctiva
        cornea
        eyelashes
        iris
        lacrimal punctae

        medial and lateral canthi
        palpebrae
        palpebral fissure
        pupil
        sclera

    B. Examine the bleached skull, wall charts, and model of eye.

        1.   What bones enter into the formation of the orbit?

          _____

          _____

        2.   What structures are found within the orbital cavity?

          _____

          _____

        3.   Review the extrinsic musculature of the eyeball, and complete the following chart:

Extrinsic Muscle	Movement of Eyeball	Cranial Nerve Supply

II. *Dissection of Fresh Beef Eyes.*

    A. Examine the specimen carefully before cutting. Locate the following structures:

        cornea   optic nerve
        conjunctiva   pupil
        extrinsic muscles   sclera
        iris

    B. With forceps (or fingers) grasp the conjunctiva near the corneal margin.

        1. Using sharp scissors cut firmly through the conjunctiva, connective tissues, and muscles down to the sclera.

        2. Continue cutting toward the back of the eyeball until you have removed all soft tissues from the surface of the sclera (except the optic nerve).

    C. Place the specimen in your watch glass, and with a scalpel make an incision in the eyeball at the corneal margin.

        1. What is the fluid which escapes? _____

        2. What is its color and consistency? _____

        _____

    D. Replace knife with scissors and continue the incision around the eyeball thus removing the entire cornea.
                 AVOID SQUEEZING THE EYEBALL during this procedure!

    E. Examine the iris which is now clearly visible. Use a blunt probe to lift the iris and produce artificial dilatation of the pupil.

        1. What is the function of the iris? _____

        _____

        2. What muscle produces:

            a. dilatation of pupil? _____

            b. constriction of pupil? _____

    F. Using scissors, enlarge your circular incision and cut away the iris. Observe the crystalline lens.

        1. How is it attached to the ciliary body? _____

        2. What is the function of the ciliaris muscle? _____

        _____

        3. Gently snip the supporting structures and remove the intact lens. Place it on a bit of newspaper.

        What does it do to the print? _____

    G. Examine the posterior portion of the eyeball.

        1. What substance fills this entire area and supports the retina? _____

2. Note the point at which the retinal blood vessels enter the eyeball. What is this general area

called? _____

_____

H. Gently remove the vitreous body and place it on a bit of newspaper.

What does it do to the print? _____

I. Using a blunt probe, nudge the soft, pinkish colored retina.

1. Does it detach easily? _____

2. At what point is it firmly anchored? _____

3. Note the relationship of the optic nerve to the front of the eyeball.

_____

J. Touch the tunic which lies just under the retina. Note the brown pigment on your finger.

1. What is this tunic called? _____

2. What is the function of the pigment? _____

_____

(Note: the greenish iridescent area or tapetum is not found in the human eye. This area reflects light rays and enables animals to see better in dim light.)

K. Observe the outer tunic of the eye.

1. What is it called? _____

2. How would you describe it? _____

_____

L. Clear your desk.

1. Wrap all tissues in newspaper and discard in appropriate waste container.

2. Wash and dry equipment before returning it to the supply table.

III. *Horizontal Section of the Eye.*

Label the following diagram as indicated. Print answers at the end of each line.

aqueous humor   optic disk
canal of Schlemm   optic nerve
choroid   retina
cornea   sclera
fovea centralis   suspensory ligament
iris   vitreous body
lens   ciliary body

Distinguish between the anterior and the posterior chambers of the eye by the use of colored pencils.

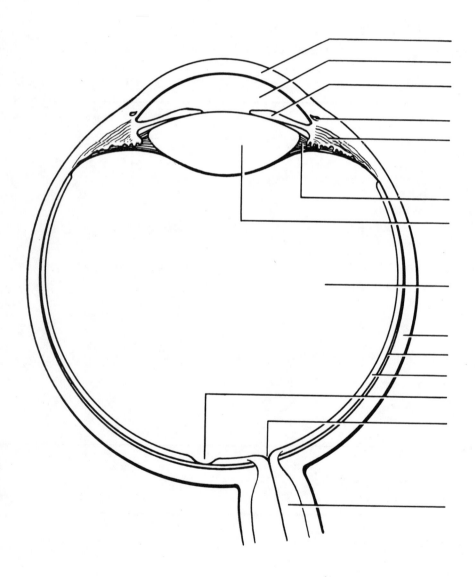

Figure 33.

IV. *Physiology of the Eye.*

   A. *Pupillary reflexes.*

      1. accommodation reflex

         a. Have your partner look at a distant object (away from the light) while you closely observe his pupils. Suddenly have him look at your finger held about 12 inches in front of his eyes.

         b. What happens to the pupils? _____

      2. light reflex. ciliospinal reflex.
         Review exercises 1 and 2 on page 74.

   B. *Refraction of light rays.*

      Three tongue blades have been placed in small beakers.
         Beaker #1 is empty
         Beaker #2 contains water
         Beaker #3 contains mineral oil

      1. Examine the beakers at eye level.

      2. Which medium produces the greatest refraction? Why?

         _____

      3. What are the major refracting media of the eye? _____

         _____

   C. *After-image.*

      1. Stare at a bright light for a minute and then look at the wall.

      2. Result? _____

   D. *Monocular vision.*

      1. Have your partner hold a test tube approximately two feet (arm's length) in front of you.

      2. Close one eye, and then quickly insert a pencil into the neck of the tube.

      3. Repeat with both eyes open.

      4. Briefly summarize your results. _____

         _____

         _____

E. *The optic disk.*

   1. Why is this area also known as the "blind spot"? _____

     _____

   2. Locating your own optic disk:

     a. Hold this page about 15 inches in front of your face so that the black cross, below, is directly in front of your left eye.

     b. Close the right eye. You should be able to see both the cross and the circle even though the left eye is focused on the cross.

     c. Move the page closer to your face until the black circle disappears.

     ⬤   ✚

     d. Explain _____

     _____

F. List, in proper sequence, the structures through which light rays and nerve impulses pass as you read this page:

   _____

   _____

V. *Application to Practical Situations.*

   A. Describe the following conditions in simple terminology.

     1. Detached retina: _____

     _____

     2. Conjunctivitis: _____

     _____

     3. Cataracts: _____

     _____

     4. Glaucoma: _____

     _____

B. What does the doctor see when he looks into a patient's eye with an ophthalmoscope? _____
_____

C. Why does a person with acute rhinitis frequently have watery eyes? _____
_____
_____

D. Why may loss of vitreous be a serious complication of eye surgery? _____
_____

E. Why do persons with one eye bandaged have difficulty in feeding themselves at first? _____
_____

F. Define the following terms:

1. Emmetropia: _____

2. Myopia: _____
_____

3. Hyperopia: _____
_____

4. Astigmatism: _____
_____

5. Nystagmus: _____
_____

G. Defects corrected by glasses. (optional)

Hold your glasses 6-8 inches away from your eyes and look through them at one of the overhead lights. Now move the glasses slowly, first to the right, then to the left. Note whether the light seems to move in the same or in the opposite direction to that in which you are moving the glasses.

1. If the light moves in the same direction the glasses have a concave lens to correct nearsightedness.

2. If the light moves in the opposite direction the glasses have a convex lens to correct farsightedness.

Hold the glasses up again, but this time as you look through them rotate the lens 90 degrees while keeping the lens at a fixed point. If the object changes in shape, a cylindrical lens has been added to correct astigmatism.

98

H. Examine the following diagram of the eyeballs, optic nerves, and optic chiasma. Assuming that a patient suffered an injury at one of the points indicated by letter A or B, what major symptoms would he present?

1. Point A: _____

_____

2. Point B: _____

_____

VISUAL FIELD

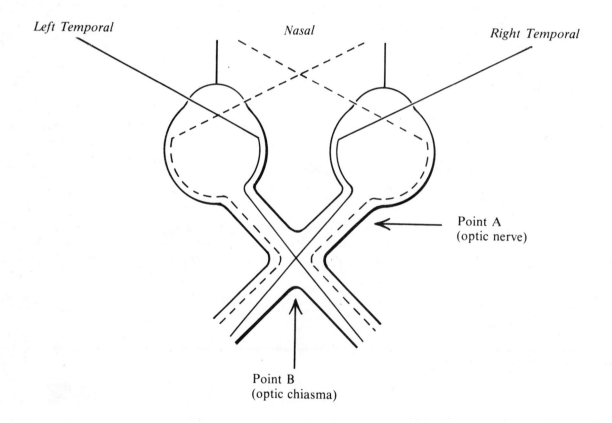

Figure 34.

Part B — **Taste Sensations**

Student Supplies:
4 test tubes on rack (for each 2 students)
prepared solutions:
    sucrose 5% (sweet)
    sodium chloride 2% (salt)
    nux vomica 2% (bitter)
    acetic acid 2% (sour)
cottton tipped applicators

phenylthiocarbamide crystals
sodium chloride crystals
ice cubes
paper cups

I. *Location of Tongue Areas for Taste.*

  A. With applicators, apply the prepared solutions to various parts of your partner's tongue. Dip applicator in solution only one time to avoid contamination.

    Rinse mouth with water between tests.

  B. Complete the following table:

Solution	Tongue Area Where Taste Most Acute	Related Cranial Nerve
sweet		
salt		
sour		
bitter		

  C. Hold an ice cube in your mouth, and then repeat the test for bitter taste.

    Result: _____

  D. Rinse mouth with warm water, and then repeat the test for bitter taste.

    Result: _____

II. *Miscellaneous Tests.*

  A. Dry your tongue with a clean paper towel. Drop several crystals of sodium chloride on the dry area. Refrain from touching tongue to other parts of the mouth.

    1. How soon could you taste the salt? _____

       _____

    2. What conclusions can you draw from this exercise? _____

       _____

B. Place a *few* crystals of phenylthiocarbamide on your tongue.

What taste sensation did you experience? _____

   (Note: Approximately 70% of a group will taste this
      substance while 30% will not.)

III. *Application to Practical Situations.*

A. Why does food seem to be relatively tasteless when one has a heavy cold? _____

_____

_____

B. List a few ways in which one could disguise the taste or odor of unpleasant oral medications. _

_____

_____

Part C — Cutaneous Sensations

Review the exercises done in Laboratory 3, page 21.

LABORATORY 13

Special Senses (continued)

Hearing and Equilibrium

Student Supplies:
anatomical wall charts
model of ear
human skulls
dissectible temporal bone

tuning forks
stop watch (or clock with loud tick)
reference books

I. *Structure of the Human Ear.*

    A. Examine your partner's external ear.

        1. What is the function of the auricle? _____

        _____

        2. Describe the curvature of the adult external auditory canal?

        _____

        3. To straighten this canal one should pull the auricle

        _____

        4. What special glands are found in the skin of this area?

        _____

        5. Their secretion is called _____

    B. Study of the middle ear. Use models and charts.

        1. What structure is found on the lateral wall of this cavity?

        _____

        2. Briefly describe the position of the ossicles and their function. _____

        _____

        _____

        3. What structure connects the middle ear with the nasopharynx?

        _____

        4. What is the function of the above structure? _____

        _____

        5. Where are the mastoid air cells located? _____

        _____

C. Study of the inner ear. Use models and charts.

   1. Membranous structures within the bony _____

      and _____

      are concerned with equilibrium, but the _____

      is concerned with hearing.

   2. What is the function of the

      a. oval window? _____

      b. round window? _____

      c. organ of Corti? _____

   3. Where do we find

      a. perilymph? _____

      b. endolymph? _____

II. *Physiology of Hearing.* (best done in a quiet room)

A. Have your partner close both eyes while seated comfortably in a chair.

   1. Stand in front of your partner, then gradually move away while holding a loudly ticking watch.

      a. How many steps did you take before the ticking could no longer be heard?

      _____

      b. Repeat while partner has fingers firmly pressed over both ears. Number of steps?

      _____

   2. Hold the ticking watch in various places within easy hearing distance. (i.e. behind the head, to the side, above the head, etc.)

      a. Where was the sound most accurately located? _____

      _____

      b. Where was it least accurately located?

      _____

      c. Explain: _____

      _____

      _____

B. List in sequence the structures through which sound waves and nerve impulses pass from the moment a word is spoken until you can understand it:

_____

_____

_____

C. Bone conduction.

   1. Strike the double end of a tuning fork on the palm of your hand to start vibrations. Hold it close to your ear until you become familiar with its sound.

   2. Plug both ears with fingers. Have partner start tuning fork vibrations, but this time place the single end against one of the bones of your skull.

   3. Can you hear the sound? _____

   4. Explain: _____

   _____

   _____

III. *Diagram of the Ear*

The following structures are illustrated on page 104.
Identify them as indicated.

acoustic nerve
auricle (pinna)
cochlea
cochlear nerve
eustachian tube
external acoustic meatus
incus

malleus
nasopharynx
semicircular canals
stapes
tympanic membrane
vestibular nerve
vestibule

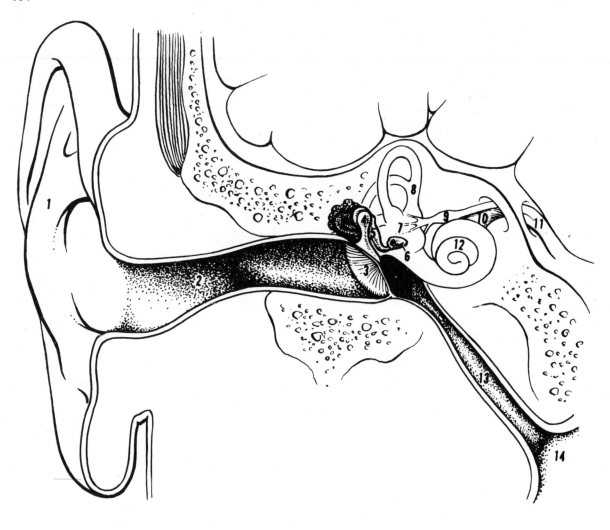

The numbers of the following spaces correspond with those in the diagram above. Write the name of each structure on the proper line.

1. _____   8. _____

2. _____   9. _____

3. _____   10. _____

4. _____   11. _____

5. _____   12. _____

6. _____   13. _____

7. _____   14. _____

Figure 35.

IV. *Maintenance of Equilibrium.*

    A. Work with your partner in testing equilibrium.

        1. Stand erect, eyes open.
            Can you stand still? _____

        2. Close both eyes.
            Can you stand without weaving slightly? _____

        3. Rotate rapidly to the right for 5 to 10 turns.

            a. Have your partner look at your eyes and describe their movements.

                _____

            b. Walk forward a few steps with eyes open and then closed. What happens?

                _____

    B. Impulses related to equilibrium are received and correlated in what part of the central nervous system?

        _____

    C. Generally speaking, where do the various afferent impulses related to the maintenance of equilibrium originate?

        _____

        _____

V. *Application to Practical Situations.*

    A. Cerumen often collects in the external auditory canal. How may it safely be removed?

        _____

        _____

    B. Why may otitis media follow a severe cold and sore throat?

        _____

        1. What is a myringotomy? _____

        2. Why is this procedure sometimes necessary in severe otitis media?

            _____

    C. What kind of deafness can be relieved by the use of a hearing aid? Why?

        _____

        _____

        _____

D. Why do we experience a sensation of pressure in the ears during sudden changes in altitude. (e.g. airplane rides)

_____

_____

E. What can be done to relieve the above sensation?

_____

F. Why may patients with injury to the dorsal columns of white matter in the spinal cord fall to the floor when they close their eyes?

_____

_____

G. What is Meniere's syndrome? _____

_____

LABORATORY 14

Introduction to the Circulatory System

Human Blood

Demonstration Supplies:
 prepared microscope slides of human blood (Wright's stain)
 microscopes
 defibrinated and clotted blood samples (optional)
 hemacytometer (optional)

Student Supplies:
 clean microscope slides and cover slips
 glass marking pencil; microscopes
 alcohol sponges (70%)
 sterile blood lancets or needles
 saline solution, 10%
 sodium citrate or oxalate solution, 0.2% to 0.3%
 Tallqvist Scale; capillary tubing
 reference books

I. *Microscopic Examination of Fresh Human Blood.*
   (Students may work in pairs or in groups of three)

 A. Select 3 clean microscope slides and number them with a glass marking pencil.

  1. Slide #1 remains dry.

  2. Place two drops of 10% saline on slide #2.

  3. Place two drops of citrate or oxalate solution at one end of slide #3.

 B. The student who is to provide the blood should wash his hands with soap and water. Then cleanse the tip of the 4th finger with an alcohol sponge and allow it to air dry.

 C. Puncture the finger with a sterile lancet or needle, and complete the preparation of slides as follows:

  1. Place one drop of blood on slide #1; using the edge of another glass slide, scrape as much blood as possible from the center of the slide. The resulting blood smear should be extremely thin and almost invisible to the naked eye.

  2. Place one drop of blood in the saline on slide #2, add a cover slip and press it down against the slide. Remove excess liquid by blotting with paper towel.

  3. Place one drop of blood on the dry end of slide #3 and a second drop of blood in the sodium citrate or oxalate solution. *Do not cover.*

 D. Examine slide #1 under the microscope. Low and high power.

  1. What type of blood cells do you see in great numbers?

   _____

  2. Are nuclei present? _____

3. How would you describe the color of these cells?

_____

4. The average number of erythrocytes ranges from _____

to _____ per cu.mm. of blood.

E. Examine slide #2. Low and high power.

1. How do these cells differ from the ones seen on slide #1?

_____

Explain: _____

_____

2. What would happen to cells placed in distilled water? _____

_____

Explain: _____

_____

F. Examine slide #3 *macroscopically.*

1. Record your observations: _____

_____

2. Explain: _____

_____

_____

3. Add cover slip if you wish to examine this slide under the microscope.

II. *Study of Prepared Microscope Slides.*

A. Examine a normal blood smear, Wright's stain, under both low and high power. Move the slide around.

1. With the aid of text and reference book illustrations identify the various types of white blood cells.

2. Distinguish between an absolute and a differential count of leukocytes.

_____

_____

_____

3. The absolute concentration of leukocytes ranges from_____

to _____ per cu.mm. of blood.

4. What is the differential concentration of leukocytes?

neutrophils   _____ %

eosinophils   _____ %

basophils   _____ %

lymphocytes   _____ %

monocytes   _____ %

5. List some of the functions of leukocytes:

_____

_____

B. Examine a blood smear, Wright's stain, taken from a patient suffering with leukemia.

How does it differ from normal blood? _____

_____

III. *Determination of Clotting Time and Hemoglobin.*

A. Another member of your team may now cleanse and puncture one finger.

B. Hemoglobin.

1. Place one or two drops of blood on a piece of filter paper.

2. As soon as the first glossy sheen disappears compare it with the Tallqvist Hemoglobin Scale. (Note: While waiting, proceed with part C.)

a. Record the percentage: _____

b. Is this method very accurate? _____

c. What is the average range in grams? _____

d. What is the function of hemoglobin? _____

_____

C. *Clotting time.*

1. Squeeze the punctured finger to insure the presence of several drops of blood.

2. Hold a 6″ to 8″ section of capillary tubing at the site so that it fills with blood. Note the time.

3. Wait for one minute and then break off a small piece of the tubing every 30 seconds until the blood "strings" showing that coagulation has occurred.

4. What was your clotting time? _____ minutes.
(include the first minute that you waited)

110

5. Distinguish between bleeding time and clotting time.

_____

_____

_____

6. What is the average range for

    a. clotting time? _____

    b. bleeding time? _____

IV. *Demonstrations.* (optional)

  A. *Defibrinated blood.*

    1. Examine the fresh clump of fibrin and the flask of blood from which it was removed.

      a. Briefly describe the appearance of the fibrin:

      _____

      b. Has the blood in the flask clotted? _____

      c. Explain: _____

      _____

    2. Observe the test tube containing a sample of blood that was permitted to stand undisturbed.

      a. Has this blood clotted? _____

      b. What is the liquid that can be seen on top of the clot? _____

      c. How does this liquid differ from blood plasma? _____

      _____

  B. *Demonstration blood count.*

  If time permits, the instructor may wish to set up the hemocytometer apparatus. Detailed instructions for this procedure are part of each kit.

V. *Application to Practical Situations.*

  A. Mr. J. B. has an extensive, suppurating infection involving his right arm.

    1. What are the symptoms of inflammation that you would expect to observe?

    _____

    _____

2. The laboratory report on a differential count of leukocytes shows a high neutrophil percentage as well as large numbers of band cells. What is the significance of this report?

_____

_____

_____

3. What is suppuration? _____

_____

B. In thrombocytopenic purpura there may be hemorrhaging from the mucous membranes.

1. Thrombocytes also are called _____

2. Normal blood shows a thrombocyte count of approximately

_____ per cu.mm.

3. What is thrombocytopenia? _____

_____

4. What part do thrombocytes play in the clotting of blood?

_____

_____

_____

5. Briefly, list the main steps in the current theory of blood coagulation.

Phase 1. _____

_____

Phase 2. _____

_____

Phase 3. _____

_____

C. Drugs of the dicumarin group (e.g. Dicumarol) tend to prevent intravascular clotting of the blood by diminishing the amount of prothrombin produced by the liver. Dosage is determined by periodic tests of the prothrombin activity in the blood or the prothrombin time.

1. What is the average prothrombin time? _____

2. Why may excessive amounts of Dicumarol be dangerous rather than helpful? _____

_____

3. What is the action of heparin in contrast to that of Dicumarol? _____

_____

112

D. What are some of the common causes of intravascular clotting of blood? _____

_____

E. Distinguish between thrombus and embolus: _____

_____

_____

F. Briefly summarize some of the advantages and disadvantages in the use of blood plasma and whole blood for intravenous administration?

     blood plasma: _____

_____

     whole blood: _____

_____

G. Mr. P. W. has experienced a chronic loss of blood over a long period of time and, as a result, is suffering from hemorrhagic anemia.

1. Generally speaking, the term "anemia" indicates a deficiency in _____

_____

2. In addition to hemorrhage, other causes of anemia may be

_____

_____

3. What is the function of erythropoietin? _____

_____

4. What stimulates the formation of erythropoietin? _____

_____

5. An increase in Mr. P. W.'s reticulocyte count will be an indication that _____

_____

H. Upon admission to the hospital, Mr. B. G. is markedly cyanotic.

1. Cyanosis is _____

2. It is due to the presence of _____

_____

3. What treatment may be ordered? _____

_____

LABORATORY 15

The Heart

Demonstration Supplies:
   fresh sheep or pig heart; fresh beef heart (optional)
   hypodermic syringe and needle
   methylene blue dye
   projection slides (optional)

Student Supplies:
   fresh sheep or pig heart, with pericardium (for each 2 students)
   dissecting trays and instruments
   small beakers or paper cups
   model of heart
   stethoscopes
   reference books
   anatomical wall charts

I.   *Dissection of Sheep Heart.*

   A.   *Preliminary examination.*

      1.   Observe the sac which surrounds the heart.

         a.   What is its name? _____

         b.   Where is it attached? _____

      2.   With scissors, make an incision in the sac near the apex of the heart. Then deliver the heart through this opening so that the sac is turned wrong-side-out and completely free except for its attachment.

         a.   What kind of membrane lines the sac? _____

         b.   Distinguish between the visceral and the parietal pericardium:

            _____

            _____

      3.   Cut away the sac but preserve the large blood vessels which enter and leave the heart.

      4.   Note the fatty streaks surrounding blood vessels on the surface of the heart.

         a.   What are these blood vessels? _____

         b.   What is their function? _____

            _____

5. Observe the general contours of the heart while holding it in the anatomic position. The latter means that

   a. the apex of the heart is directed _____

   _____

   b. the base of the heart is directed _____

   _____

6. Identify the two ear-like (auricular) appendages at the base of the heart. These lie over the upper chambers (atria) of the heart.

7. With thumb and forefinger squeeze the wall of the lower portion of the heart. One side will feel much thicker than the other.

   a. The thicker side is the _____ ventricle.

   b. The thinner side is the _____ ventricle.

B. *Study of the right heart.*

1. Locate the large veins which open into the right atrium. They are thin-walled, collapsed vessels near the right auricular appendage and similar in color to the appendage.

   a. What are their names? _____

   b. What is their function? _____

   _____

2. Insert a blunt probe into these vessels, then cut along the line of your probe so that the interior of the right atrium may be clearly visualized. (Note: if large blood clots are present they should be removed with your forceps.)

3. Examine the atrium very carefully and locate the depressed, white scar on the interatrial septum.

   a. What is this scar called? _____

   b. What fetal structure did this scar replace and what was its function? _____

   _____

4. In addition to the venae cavae there is another vascular structure opening into the right atrium. It can easily be located just below the fossa ovalis.

   a. What is this structure? _____

   b. What is its function? _____

   _____

5. Pour a small amount of water through the atrium and into the ventricle. Gently squeeze the ventricle and note the action of the atrioventricular valve. (Note: the instructor may demonstrate this action in a beef heart.)

    a. What is another name for this valve? _____

    b. What is its function? _____

    _____

6. With scissors, cut down through the wall of the right ventricle so that the entire inner aspect is clearly visible.

    a. The lining of the heart is called _____

7. Observe the muscular columns which give rise to slender tendinous strings. Use your blunt probe to lift these strings and determine their superior point of attachment.

    a. The strings are called _____

    b. They extend from the _____ muscle

    columns to the flaps of the _____ valve.

    c. Their function is _____

    _____

    d. How many flaps (cusps) does this valve have? _____

8. Locate the interventricular septum. Then run your index finger upward along the septum, under the valve flaps until it enters the blood vessel leaving the right ventricle.

    a. What is this vessel? _____

    b. What is its function? _____

    _____

9. With scissors, open this vessel and closely examine the proximal portion.

    a. What valve flaps do you see? _____

    b. What is the function of this valve? _____

    _____

C. *Study of the left heart.*

    1. Examine the upper surface of the left atrium.

        a. What blood vessels open into this chamber?

        _____

        b. What is their function? _____

        _____

2. With scissors, cut into the atrium and then continue your incision down through the wall of the ventricle to the apex.

   a. How does the left atrioventricular valve differ from the right?

   _____

   _____

   b. What other names are given to this valve? _____

   _____

   c. What is its function? _____

   _____

3. Compare the right and left ventricular walls.

   a. Which are thicker? _____

   b. What is a logical explanation for this difference?

   _____

   c. Heart muscle also is called _____

4. Run your index finger up under the valve flaps near the septum until you enter the large blood vessel leaving the left ventricle.

   a. What is the name of this vessel? _____

   b. What is its function? _____

   _____

5. Open this blood vessel with your scissors and closely observe the proximal portion.

   a. What valve flaps do you see? _____

   b. What is their function? _____

   _____

   c. What are the two small openings just above the valve flaps? _____

   d. These vessels convey blood to the _____

   _____

D. Wrap hearts in newspaper before discarding in appropriate waste container. Clean and replace equipment.

II. *The Coronary Blood Vessels.*

    A. If time permits, the instructor may inject methylene blue dye into one of the coronary blood vessels of a fresh heart for demonstration purposes.

    B. List, in sequence, the vessels through which a drop of blood will pass as it travels from the aorta to the myocardium and then to the right atrium:

aorta

    C. Why is the coronary circulation of such vital importance?

_____

_____

_____

III. *Chambers and Valves of the Heart.* Diagrammatic

    A. Label the following diagram as indicated:

aorta	pericardium
aortic semilunar valve	pulmonary artery
bicuspid or mitral valve	pulmonary vein
chordae tendineae	pulmonary semilunar valve
inferior vena cava	right atrium
left atrium	right ventricle
left ventricle	superior vena cava
myocardium	tricuspid valve
papillary muscle	

    B. With blue pencil shade those areas which contain blood low in oxygen. Use red pencil to shade areas containing blood that is rich in oxygen.

    C. Add arrows to indicate the direction of blood flow through the heart.

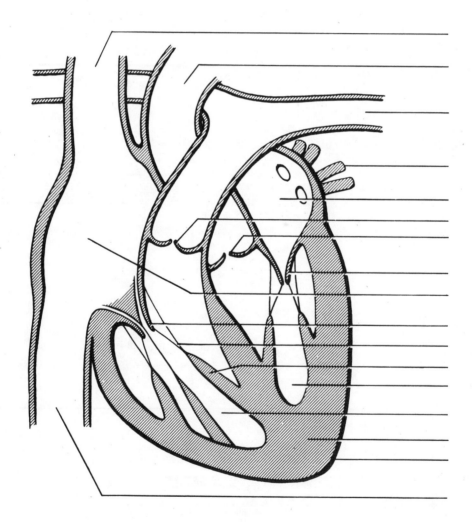

Figure 36.

IV. *The Functioning Heart.*

    A. *Heart sounds.*

        1. Working with your partner, locate the chest area that is over the apex of the heart. Clothing should be pulled to one side if possible.

            a. What landmarks are used in locating this area?

            _____

            b. Can you feel the pulsation of the heart with your finger tips?

        2. Place the bell of your stethoscope over the apex of your partner's heart. Listen carefully to the sounds.

            a. How would you describe them? _____

            _____

            b. What causes these sounds? _____

            _____

            c. Briefly, what is meant by a "heart murmur"?

            _____

            _____

        3. Apical-radial pulse determination.

        While seated comfortably at a table, listen to your partner's apical heart beat. Your partner, in turn, will count his/her own radial pulse. Place a stop watch where both can see it and count beats for one minute. Start counting together on prearranged signal. Record your results below:

            _____ apical pulse

            _____ radial pulse

            _____ pulse deficit (if any)

    B. *The wave of contraction.*

        1. This wave is initiated by the _____ or pacemaker of the heart and sweeps over the atria.

        2. How is the wave transmitted to ventricular muscle tissue which is not continuous with that of the

        atria? _____

        _____

        _____

3. What is atrioventricular heart block? _____

_____

    a. What happens to atrial contractions? _____

    _____

    b. What happens to ventricular contractions? _____

    _____

    c. Why does the radial pulse rate drop to 30-35 per minute?

    _____

    _____

    d. Is this compatible with life? _____

C. *The cardiac cycle.*

1. This cycle includes _____ (the period of contraction), and _____ (the period of dilatation and rest).

2. The average time required for one complete cycle is _____ second.

3. When the heart beats more rapidly, as in exercise, it is the period of _____ that is shortened.

V. *Application to Practical Situations.*

A. You have often heard that in coronary thrombosis with myocardial infarction complete bed rest is ordered.

1. What is coronary thrombosis? _____

_____

_____

2. Myocardial infarction refers to _____

_____

_____

3. Why is bed rest of such great importance? _____

_____

_____

_____

4. Many persons survive a heart attack, but coronary thrombosis *can* cause instantaneous death.

   Why? _____

   _____

   _____

B. Mr. G. R. has organic heart disease with severe atrial *fibrillation,* one of the most common chronic irregularities of the heart. His apical-radial pulse is 150-90.

   1. What is atrial fibrillation? _____

   _____

   _____

   2. What effect does it have on the ventricular beat? _____

   _____

   _____

   3. If Mr. G. R.'s apical-radial pulse is 150-90 what is his pulse deficit? _____

   4. Why is the radial pulse less than the apical pulse?

   _____

   _____

   _____

C. Briefly, explain the following terms:

   1. endocarditis: _____

   _____

   2. mitral stenosis: _____

   _____

   3. pericarditis: _____

   _____

   4. tachycardia: _____

   _____

   5. bradycardia: _____

   _____

   6. aortic insufficiency: _____

   _____

D. *The Normal Electrocardiogram.*

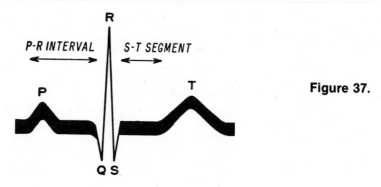

Figure 37.

1. An electrocardiogram provides information regarding

_____

2. What does the P wave represent? _____

_____

3. The P - R interval is _____

_____

4. What does the QRS wave represent? _____

_____

5. The S - T segment represents _____

_____

6. What does the T wave represent? _____

_____

LABORATORY 16

Major Arteries of the Body

Demonstration Supplies: (Optional)
    dissected embalmed cat
    projection slides
    prepared microscope slides

Student Supplies:
    anatomical wall charts
    reference books

I.   *A Study of the Structure of the Arteries.*

    A. Complete the following statements:

        1. Arteries convey blood from the _____

           to the _____

        2. Arteries are composed of three coats or tunics. Nourishment of tissue cells in these tunics

           depends on networks of tiny blood vessels that are called the _____

           _____

        3. Why are extensibility and elasticity such important characteristics of the arteries? _____

           _____

           _____

        4. When an artery is severed, the orifice remains open and blood flows in great _____

        5. How do the elastic fibers and muscular coat of an artery function in the arrest of hemorrhage?

           _____

           _____

           _____

        6. What is the general function of

           a. the pulmonary arteries? _____

           _____

           b. the systemic arteries? _____

           _____

124

II. *Diagrams of Major Arterial Pathways.*

Label the following diagrams as indicated. *Print* at the end of each label line.

A. *The Circle of Willis.*

anterior cerebral
anterior communicating
basilar
internal carotid

middle cerebral
posterior communicating
posterior cerebral
vertebral

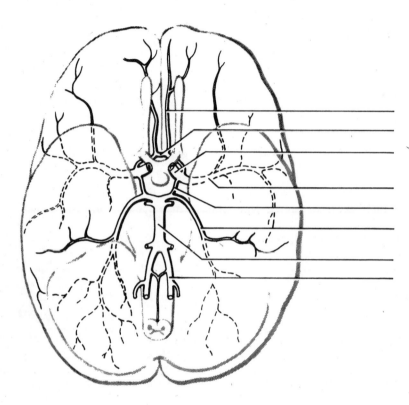

Figure 38.

1. List, in sequence, the vessels through which a drop of blood would pass as it traveled from the arch of the aorta to capillaries in the frontal lobe of the brain.

aortic arch ⟶

2. Trace a second drop of blood from the subclavian artery to capillaries in the occipital lobe of the brain.

subclavian artery ⟶

B. *Major Arteries of Thorax and Abdomen.*
   (Exclusive of Celiac Axis)

aortic arch
common carotid
common iliac
external iliac
inferior mesenteric
brachiocephalic (innominate)
intercostal
internal iliac

phrenic (inferior)
pulmonary (left)
renal
spermatic (or ovarian)
subclavian
superior mesenteric
suprarenal (middle)
vertebral

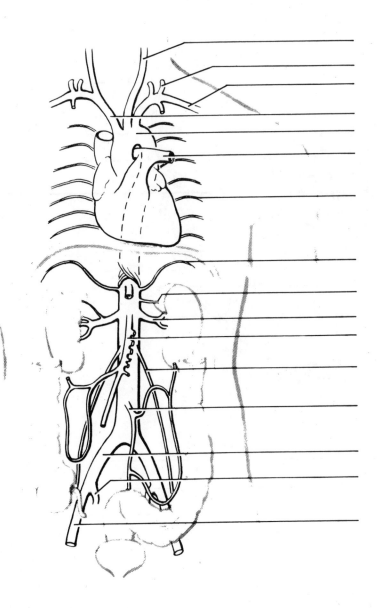

Figure 39.

C. *The Celiac Axis.*

aorta
celiac
cystic
duodenal
gastroduodenal
hepatic

left gastric
left gastroepiploic
pancreatic
right gastric
right gastroepiploic
splenic

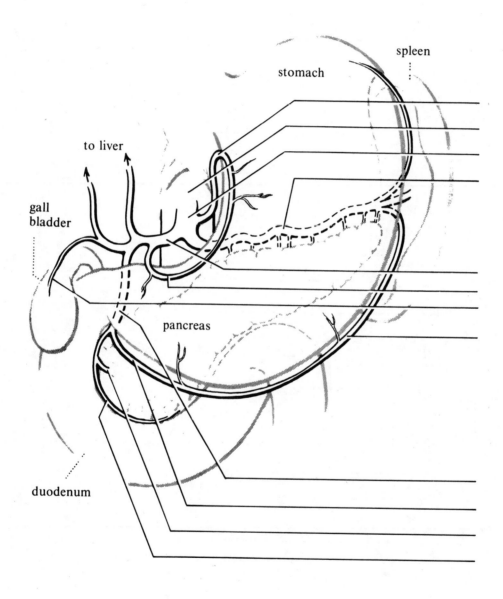

spleen

stomach

to liver

gall
bladder

pancreas

duodenum

Figure 40.

D. *Major Arteries of the Extremities.*

anterior tibial
axillary
brachial
digital
dorsalis pedis
external iliac
femoral

popliteal
posterior tibial
radial
subclavian
ulnar
volar arch (deep)
volar arch (superficial)

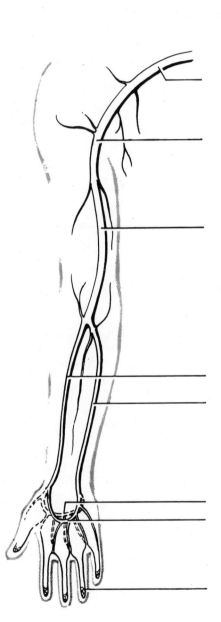

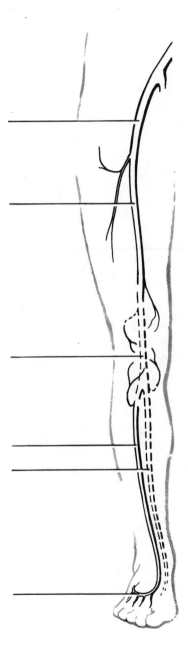

Figure 41.

III. *Demonstration of the Cat.* (optional)

The instructor may point out the major arteries as they are found in the dissected, embalmed cat or this demonstration may be delayed until the veins can be included.

IV. *Problems.*

    A. Working with your partner, attempt to palpate the various arteries as they approach the surface of the body.

        Record names of arteries and points at which they were palpated:

              Artery   *Point of Palpation*

    B. List, in sequence, the vessels through which a drop of blood must pass as it travels from the aortic arch to the top of your left foot.

            aortic arch

               ↓

C. Trace the pathway of a second drop of blood from the aortic arch to the palm of the right hand.

aortic arch
↓

D. Trace the pathway of a third drop of blood from the aortic arch to the spleen.

aortic arch
↓

V. *Application to Practical Situations.*

A. You learn that a friend of yours has an aortic aneurysm. He does not seem to be in any great discomfort, but his name has been included in the list of critically ill patients.

1. What is an aneurysm? _____

_____

2. Why is your friend listed as a critically ill patient?

_____

_____

B. You sometimes hear or read the phrase "hardening of the arteries."

1. What is the medical term for this condition? _____

_____

2. Briefly, what is happening to the arteries? _____

_____

C. Mr. S. C. has Buerger's disease (thromboangiitis obliterans) with a secondary gangrene of the left foot.

1. Briefly, what is Buerger's disease? _____

_____

2. What is gangrene and why may it follow as a complication of Buerger's disease? _____

_____

_____

_____

_____

3. Why must these patients give up the smoking of cigarettes?

_____

_____

4. The doctor has ordered daily palpation of Mr. S. C.'s right dorsal pedal pulse. Why? ____

_____

How would you do this? _____

_____

Check the dorsal pedal pulses of your partner.

LABORATORY 17

The Return of Blood and Lymph

Part A — **Major Veins of the Body**

Demonstration Supplies:
   dissected embalmed cat
   prepared microscope slides
   projection slides

Student Supplies:
   anatomical wall charts
   reference books

I. *A Study of the Structure of Veins.*

   A. Complete the following statements:

      1. Veins convey blood from the _____ to

          the _____

      2. Structurally speaking, what are the main differences between veins and arteries? _____

          _____

          _____

          _____

      3. Briefly, describe the structure of the valves which are found in many of the veins: _____

          _____

          _____

      4. What is the function of these valves? _____

          _____

      5. What happens when a vein is severed? _____

          _____

   B. Examine the demonstration microscope slide showing an artery, vein, and nerve in cross-section. Make a rough sketch of your observations.

II. *Demonstration of the Cat.*
   The instructor will demonstrate anatomical features of the circulatory system in an embalmed cat. A doubly or triply injected specimen may be used.

III. *Diagrams of Major Venous Pathways.*

Label the following diagrams as indicated. *Print* at the end of each label line.

A. *Cranial Venous Sinuses.*

cavernous sinus
falx cerebri
inferior sagittal sinus
internal jugular vein

straight sinus
superior sagittal sinus
tentorium cerebelli
transverse (lateral) sinus

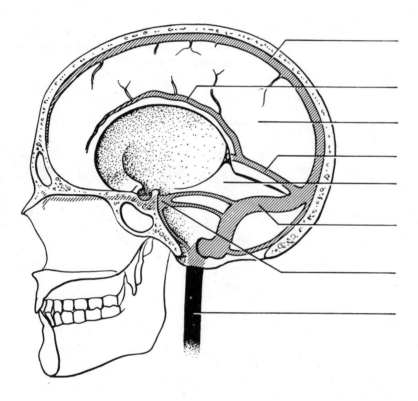

Figure 42

133

B. *Major Veins of the Thorax and Abdomen:*
(Exclusive of Portal System)

azygos
brachiocephalic (innominate)
common iliac
external jugular
external iliac
hemiazygos
hepatic
inferior vena cava

intercostal
internal iliac
internal jugular
renal
spermatic (or ovarian)
subclavian
superior vena cava
suprarenal

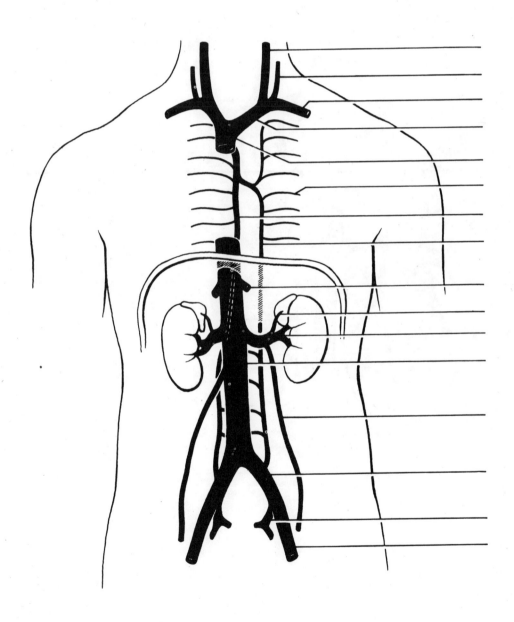

Figure 43

C. *The Portal System of Veins.*

cystic
duodenal
gastric
hemorrhoidal (rectal)
inferior mesenteric

pancreatic
portal
splenic
superior mesenteric

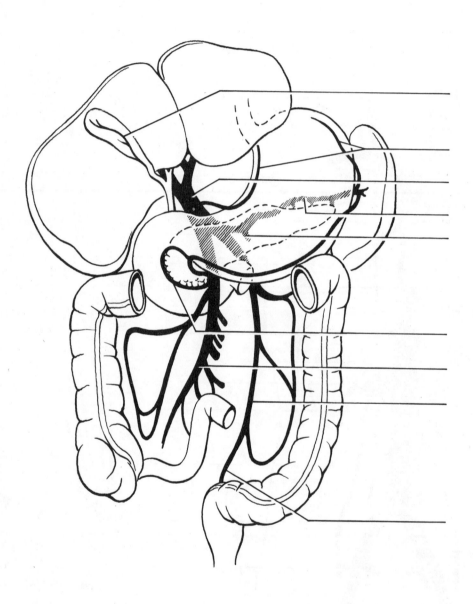

Figure 44.

D. *Major Veins of the Upper Extremity.*

Superficial Set

basilic
cephalic
median
median cubital

Deep Set

axillary
brachial
radial
subclavian
ulnar
volar arches

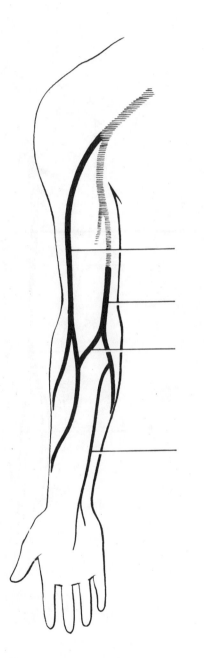
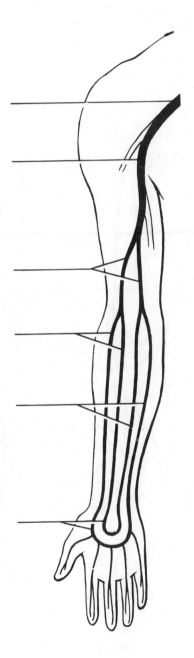

Figure 45.

E. *Major Veins of the Lower Extremity.*

    Superficial Set   *Deep Set*

        great saphenous   anterior tibial
        small saphenous   femoral
                                    popliteal
                                    posterior tibial

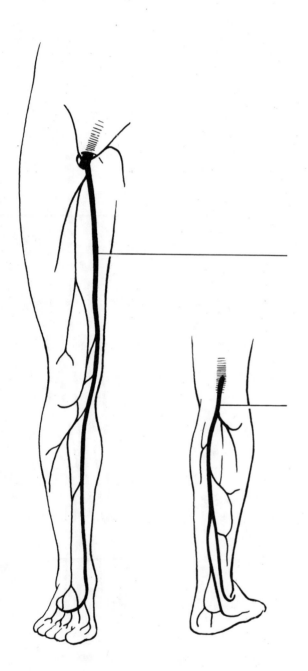
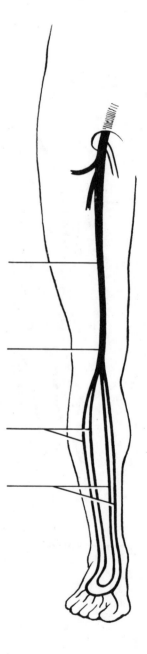

Figure 46.

IV. *Application to Practical Situations.*

    A. Mrs. B. B. has severe varicose veins. Her doctor plans to ligate (tie off) and remove both of the great saphenous veins.

        1. Briefly, what are varicose veins? _____

        _____

        2. They are commonly caused by _____

        _____

        3. People in what occupations are particularly prone to develop varicose veins? _____

        _____

        4. If the great saphenous vein is removed how will blood be returned from superficial areas of the leg and thigh? _____

        _____

        5. What other superficial vein of the lower extremity may develop varicosities? _____

        _____

        6. What might be done to help prevent the development of varicose veins? _____

        _____

    B. A severe phlebitis of the ilio-femoral veins may be associated with several pulmonary emboli.

        1. What is phlebitis? _____

        2. What is pulmonary embolism and why is it a serious complication of phlebitis? _____

        _____

        _____

        _____

        3. What drugs might be given to prolong the clotting time? _____

        _____

    C. Portal hypertension may be treated surgically by portacaval shunt, in which the portal vein is anastomosed with the inferior vena cava.

        1. What is the general function of the portal system of veins? _____

        _____

        2. What is portal hypertension? _____

        _____

3. Why would a portacaval shunt be of value? _____

_____

_____

4. Complete the following table:

Structure Drained	By What Vein?	Which Terminates In
stomach	_____	_____
spleen	_____	_____
pancreas	_____	_____
small intestine, cecum, asc. and trans. colon	_____	_____
rectum, sigmoid and descending colon........	_____	_____
gallbladder	_____	_____
liver	_____	_____

Part B — The Lymphatic System

Demonstration Supplies:
    dissected embalmed cat, lymphatic injection (optional)
    projection slides (optional)
    prepared microscope slides

Student Supplies:
    anatomical wall charts
    reference books

I. *A Study of the Structure and Function of Lymphatics.*

A. Examine the prepared microscope slides of lymph nodes, spleen, and collecting vessels. Compare with textbook pictures.

B. Demonstration of specially injected cat. (optional)

The instructor may point out spleen, lymph nodes, Peyer's patches, collecting vessels, and terminal ducts in a cat when such specially injected animals are available.

C. With the aid of text and reference books, answer the following questions:

1. Briefly, describe interstitial fluid on the basis of its origin and composition:

_____

_____

2. How is interstitial fluid returned to the bloodstream? _____

_____

_____

3. How do lymph capillaries differ from blood capillaries? _____

_____

4. What are the functions of lymph nodes? _____

_____

5. What outstanding structural differences can be noted between lymph collecting vessels and veins
of the blood vascular system? _____

_____

_____

6. The onward flow of lymph depends on _____

_____

7. Where is the spleen located? _____

_____

    a. What are some of its functions? _____

    _____

    _____

    b. Which of these functions could be related to the sharp pain that may be felt in the left side
    during sudden physical exertion? Explain.

    _____

    _____

II. *Diagram of Lymph Drainage Patterns.*

Complete the diagram on page 140 as indicated.

A. With colored pencils lightly shade those portions of the body which are drained by the right lymphatic duct and by the thoracic duct.

B. Sketch in a rough diagram of the general location of the aforementioned terminal ducts.

C. Add small "x" marks, to indicate those areas where lymph nodes are especially abundant.

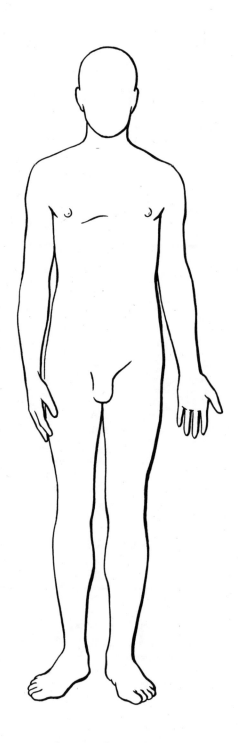

Figure 47.

III. *Application to Practical Situations.*

    A. Mr. B. L. has a severe infection in his left hand accompanied by red streaks in the skin of his forearm.

        1. What is responsible for the red streaks? _____

           _____

           _____

        2. Why is this a dangerous development? _____

           _____

           _____

    B. A newspaper article mentioned the phrase "elephantiasis of the right leg."

        1. What is elephantiasis? _____

           _____

           _____

    C. What is edema or dropsy? _____

        _____

    D. List, in sequence, the main structures through which a drop of chyle would pass as it travels from the intestine to the bloodstream:

LABORATORY 18

The Physiology of Circulation

Demonstration Supplies:
> live frog, 10% urethane solution
> cork board with 3/4″ hole at one end
> dissecting set, microscope
> hypodermic syringe and needle
> sphygmomanometer, stethoscope
> amyl nitrite aspiroles
> motion picture film
> projection slides (optional)

Student Supplies:
> live frog for each 2 or 3 students
> dissecting trays and instruments
> cork board with 3/4″ hole at one end
> microscopes; absorbent cotton
> common pins; ice

I.   *A Study of Circulation in the Frog.*

A.   The instructor will inject 1 cc. of 10% urethane solution into the ventral lymph sac or into the peritoneal cavity of each frog. This produces general anesthesia in about 5 minutes. When frog is quiet proceed with the following work.

B.   Place frog *dorsal side up* on a cork board.

1.   Open its mouth and pull the tongue out. *Note:* the frog's tongue is attached anteriorly.

2.   Using no more than 4 pins, secure the tongue over the hole in the cork board so that light can shine through it when under the microscope. Do not stretch tongue too tightly or circulation will be obstructed.

3.   Examine this preparation under *low power.* Do not permit lens to touch the tongue.

4.   Distinguish between arterioles and venules on the basis of the direction of blood flow.

a.   from larger vessels into smaller = _____

b.   from small vessels into larger = _____

5.   Observe the capillary vessels.

a.   How does the rate of flow compare with that in arterioles and venules? _____

_____

b.   Why is this difference in rate of flow essential?

_____

   c. How does the diameter of the capillary compare with that of the erythrocytes passing through?

_____

_____

   d. Note the flexibility and elasticity of the erythrocytes as they pass especially narrow portions of the capillary.

C. Remove frog from board and place it ventral side up in a dissecting tray.

D. With scissors, open the chest cavity and expose the heart.

   1. Record the rate: _____ beats per minute.

   2. Place a small piece of ice directly on the heart for one minute. What happens to the rate?

        _____ beats per minute.

   3. Cover the heart with a cotton ball that has been dipped in warm water (approximately 45 C.). What happens to the rate?

        _____ beats per minute.

   4. Cut a large artery as it leaves the heart. What happens to the rate? (Note: Omit this experiment if there has been excessive blood loss during opening of chest cavity).

        _____ beats per minute.

E. Wrap frogs in newspaper before discarding in appropriate container. Clean and replace equipment.

F. Demonstration of inflammatory processes. (optional)

   1. The instructor may make a small incision in the lateral abdominal wall of an anesthetized frog. With forceps the intestinal loops are gently pulled from the body and pinned over the hole in a cork board so that light shines through the mesentery. This preparation is then placed under the demonstration microscope.

   2. Students should observe the circulation of blood in the mesentery at 30 minute intervals during the laboratory period.

   3. Note any changes in the size of the vessels and in the rate of blood flow: _____

_____

_____

II. *A Study of Pulse Rate and Blood Pressure.*
(Note: The following exercises may be done as a demonstration by the instructor and student volunteers unless time and equipment permit the entire laboratory section to participate, possibly in groups of 3 or 4).

A. *The effects of exercise.*

   1. Subject #1 sitting quietly in a chair.

      a. blood pressure: _____

      b. pulse rate: _____

2. Subject exercises by running in place for one minute. (leave blood pressure cuff on arm)

    a. blood pressure: _____

    b. pulse rate: _____

3. What are the effects of exercise on blood pressure and pulse rate? Explain briefly.

    _____

    _____

    _____

    _____

B. *The inhalation of amyl nitrite.* (optional)

    1. Subject #2 sitting quietly in a chair.

        a. blood pressure: _____

        b. pulse rate: _____

    2. Crush amyl nitrite aspirol and allow subject to take 2 or 3 inhalations. *Note:* take blood pressure immediately! Pulse may be taken by a student volunteer.

        a. blood pressure: _____

        b. pulse rate: _____

        c. appearance of skin: _____

    3. What is the action of this drug? _____

        _____

    4. Have subject describe the sensations that were experienced after inhaling this drug: _____

        _____

        _____

III. *Answer the Following Questions.*

    A. Distinguish between arteries, veins, and capillaries by checking the appropriate spaces below:

	Velocity of Blood Flow			Blood Pressure		
	rapid	medium	slow	high	medium	low
arteries						
capillaries						
veins						

B. What is meant by the terms "blood pressure" and "pulse pressure"?

_____

_____

_____

    1. What are two major factors that affect the pulse pressure?

    _____

    _____

    2. What is the average range for arterial blood pressure? _____

    _____

    3. List the major factors involved in the maintenance of arterial blood pressure:

    _____

    _____

    _____

    _____

C. The pulse is caused by _____

_____

    1. The average adult pulse rate is _____ per minute.

    2. A sudden rise in arterial blood pressure causes a reflex _____ in the pulse rate, but a

    sudden drop in arterial blood pressure causes a reflex _____ in pulse rate.

    3. A marked increase in the amount of blood entering the right atrium of the heart will have what effect on the pulse rate? Explain.

    _____

    _____

    _____

    _____

IV. *Application to Practical Situations.*

    A. Generally speaking, what would happen to the heart rate in each of the following instances?

        1. temperature elevation (fever): _____

        2. hemorrhage (severe): _____

        3. hypothermia (subnormal temperature): _____

B. What effect does congestive heart failure have on the venous blood pressure? Explain briefly.

_____

_____

_____

   1. A compensated heart is _____

      _____

   2. A decompensated heart is _____

      _____

C. Miss R. M. has occasional attacks of *angina pectoris* following undue exertion. She carries nitroglycerin tablets in her purse for emergency relief of pain.

   1. What is angina pectoris? _____

      _____

      _____

   2. Why is it relieved by nitroglycerin? _____

      _____

D. A patient who had been lying flat in bed for several days was told that he could begin to resume regular activity. Upon assuming a standing position, however, he complained of feeling rather faint. In a few seconds the sensation of faintness disappeared, but his pulse was extremely rapid.

   1. Why did he feel faint? _____

      _____

   2. Explain, in terms of protective reflexes, the significance of the rapid pulse rate:

      _____

      _____

      _____

E. A neighbor described her illness as primary or essential hypertension.

   1. An individual is said to have hypertension when _____

      _____

   2. Essential hypertension is _____

      _____

   3. What is secondary hypertension? _____

      _____

LABORATORY 19

The Respiratory System

Part A — **Structural Components**

Demonstration Supplies:
  fresh sheep lungs with trachea
  glass tubing, one-hole rubber stopper
  dissecting tray, hemostats
  model of larynx; dissectible skull
  human torso model; wall charts
  dissected embalmed cat
  prepared microscope slides

Student Supplies:
  set of fresh sheep lungs with trachea and larynx for each 4 students
  dissecting trays, scissors, beaker of water
  text and reference books

I. *Review of the Paranasal Sinuses.*

   A. Examine the dissectible skull and identify the paranasal sinuses.

      1. What are their names? _____

      _____

      2. They are lined with _____membrane, and they open into the

      _____

   B. What are the functions of the paranasal sinuses? _____

   _____

II. *Dissection of Respiratory Organs of the Sheep.*

   A. *Examination of the Larynx.*

      1. Observe carefully the gross structure. Note openings into larynx and esophagus.

      2. With scissors, remove esophagus and the musculature surrounding the larynx.

      3. Identify the major cartilages which have been exposed.

         a. What structure prevents food from entering the larynx?

         _____

         b. The laryngeal prominence (Adam's apple) is formed by the _____ cartilage.

         c. Which cartilage in shaped like a signet ring?

         _____

4. With scissors, cut through the posterior wall so that the interior of the larynx may be clearly visualized. (Note: the sheep does not have vocal folds, or cords, comparable to those in the human.)

   a. Briefly differentiate between vocal folds and ventricular folds:

_____

_____

_____

   b. The opening between the vocal folds is known as the _____

_____and is the narrowest part of the respiratory tract.

B. *Examination of the trachea.*

1. Feel the cartilaginous rings and note the U-shaped appearance. Pull on the trachea and note its elasticity.

2. With scissors, make a mid-line incision down the posterior aspect of the trachea to its bifurcation.

3. Feel the lining of the trachea.

4. What is the function of the cartilaginous rings? _____

_____

5. What fills in the posterior aspect or open portion of the "U"? _____

6. The trachea is lined with _____membrane.

The top layer of this membrane is composed of _____ epithelium.

7. The trachea is located _____ to the esophagus.

C. *Examination of the bronchi and lungs.*

1. Observe the general contour, color, and consistency of fresh lung tissue.

   a. Note the number of lobes and compare with human lungs:

	Human	Sheep
Right lung	_____	_____
Left lung	_____	_____

   b. The lungs are covered with a serous membrane which is called the _____

_____

c. The visceral layer of this membrane is closely applied to _____ ,

_____

while the parietal layer is applied to _____

_____

d. The potential space between these two layers of membrane is called _____

c. What is the color of fresh sheep lung? _____

f. People who live in a large city usually have lungs that are _____ in color.

g. Describe the consistency of fresh lung tissue:

_____

2. With scissors, continue your tracheal incision down into the bronchi, branching right and left. Note the many ramifications and feel the walls. Pinch the lung tissue between your fingers.

a. Briefly, describe the cartilaginous structures of bronchi and bronchioles in contrast to that of the trachea:

_____

_____

b. When foreign bodies are aspirated through the glottis they are most often found in the right bronchus.

Why? _____

_____

3. Snip off a portion of lung tissue and drop it in a beaker of water. Why does lung tissue float even

when cut into small sections? _____

_____

D. Wrap lungs in newspaper before discarding in appropriate waste container. Clean and replace equipment.

III. *Demonstration Lung Inflation.* (instructor)

Insert a section of glass tubing into a small, one-hole rubber stopper. Place stopper in sheep trachea where it may be tied with cord or held firmly with one hand. Blow into the tubing to inflate the lungs.

Note: lungs will inflate more readily if their surface is kept moist and if the first few inflations are accompanied by a gentle, kneading massage to facilitate the opening of collapsed bronchioles. Hemostats may be used to control leakage from any small defects.

IV. *Respiratory System of the Cat.* (optional)

Students may examine the larynx, trachea, lungs, bronchial ramifications, and pulmonary blood vessels in the dissected embalmed cat. Note the position and attachment of the diaphragm.

V. *Additional Visual Aids.*

    A. Examine the respiratory organs in the dissectible human torso model.

    B. Look at the prepared microscope slides of human lung tissue. Compare with illustrations in your textbook.

    C. Examine the model larynx.

    D. Study the anatomical wall charts.

    E. The instructor may show a motion picture film.

VI. *Diagrams of Human Respiratory Structures.*

Label the following diagrams as indicated. *Print* at the end of each label line.

A. *The Upper Respiratory System.*

epiglottis

frontal sinus

larynx

nasal conchae

nasopharynx

oropharynx

pharyngeal tonsil (adenoid)

sphenoid sinus

trachea

vocal fold (cord)

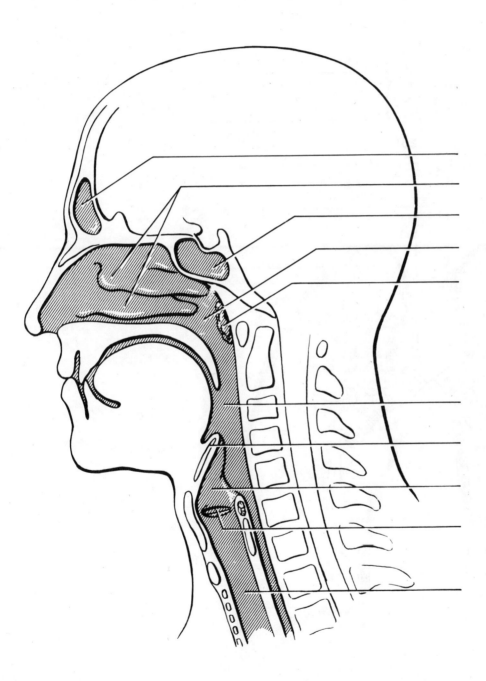

Figure 48.

154

B *Lower Respiratory System.*

apex of lung   mediastinum
base of lung   parietal pleura
bronchiole   pleural cavity
cricoid cartilage   right bronchus
diaphragm   thyroid cartilage
epiglottis   trachea
left bronchus   visceral pleura

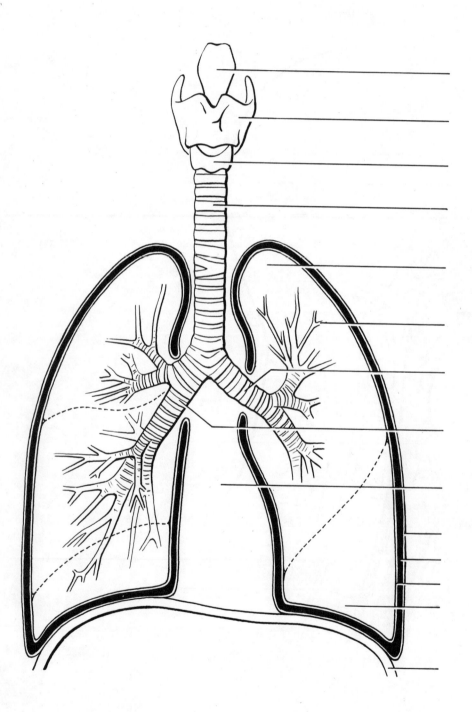

Figure 49.

C. *Pulmonary Lobule.*

alveolar ducts
alveolar sacs
alveoli
respiratory bronchiole
terminal bronchiole

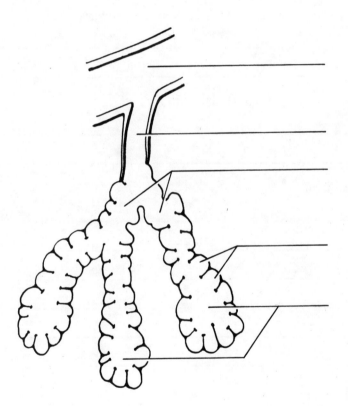

Figure 50.

VII *Application to Practical Situations.*

A. Teddy was playing with a small whistle which he accidentally aspirated. Within minutes he was extremely cyanotic, and a doctor who lived nearby performed an emergency, life-saving tracheostomy. The whistle was later removed at the hospital, and the tracheostomy was closed.

1. Where did the whistle get caught? _____

_____

2. What is cyanosis? _____

_____

3. What is tracheostomy and why did it save Teddy's life?

_____

_____

4. Why are tracheostomy patients particularly susceptible to pulmonary complications?

_____

_____

5. Could Teddy talk while this opening was in his trachea?

Explain _____

_____

B. Briefly, explain the following medical terms:

1. pleurisy _____

2. pharyngitis _____

3. bronchitis _____

4. pneumonia _____

5. pneumonectomy _____

6. lobectomy _____

7. bronchoscopy _____

8. thoracotomy _____

9. laryngectomy _____

10. pulmonary emphysema _____

_____

Part B — **Physiology of Respiration**

Demonstration Supplies:
    live frog; dissecting instruments
    glass slide and cover slip
    microscope

Student Supplies:
    live frog for each 3 or 4 students
    blunt probes, scissors, cork boards
    common pins; bit of plain cork (1 mm.)
    cotton; ice; ether; Ringer's solution
    tape measures; stethoscopes
    brown paper bags; vital capacity machine
    bell jar with balloons and rubber diaphragm
    text and reference books

I.   *A Study of Ciliated Membrane in the Frog.*

    A.   *Preparation of the frog.* (brain and cord pithed by instructor)

        1.   Place frog ventral side up on cork board.

        2.   With heavy scissors, open abdominal and chest cavities from pubis to lower jaw.

        3.   Pin forelegs to cork board. Remove heart and liver.

        4.   Insert a blunt probe into frog's mouth and gently push probe down the esophagus into the stomach.

        5.   Insert scissors and cut along the line of your probe thus opening the lower jaw, esophagus, and stomach in a midsagittal plane.

        6.   Hold these structures in a straight line by inserting a pin in the stomach and anchoring it to the board.

        7.   Use cotton sponges dipped in Ringer's solution to remove excess mucus and to moisten surface of the membrane at frequent intervals.

        8.   Measure off a distance of approximately ½″ on the surface of the membrane and mark it with two pins.

    B.   *Observation of cilia in action.*

        1.   Moisten a small piece of cork and place it at the upper end of your measured course. Note the time required for it to move to the other end.

           Record time here: _____

        2.   Repeat this procedure while tilting the board so that the frog's feet are higher than its head. (30° angle)

           Did the pull of gravity cause any appreciable difference in the time required for the cork to travel the course?

           _____

3. Place some ice on the membrane for one minute. Remove ice and excess fluid. Repeat the cork procedure. What happened to the rate of movement?

_____

4. Cover the membrane with cotton which is dripping wet with warm Ringer's solution (40-45°C.) Remove cotton and excess fluid after one minute. Repeat the cork procedure. What happened to the rate of movement?

_____

5. Flood the membrane with several drops of ether. Repeat the cork procedure. What happened to the rate of movement?

   Explain. _____

   _____

C. *Microscopic appearance of ciliated epithelium.* (demonstration)

Examine the demonstration microscope preparation of living ciliated epithelium under both low and high power. Focus on the edge of the material for best results, and reduce light with aid of the iris diaphragm.

This preparation is made by removing a small section of membrane from the frog's throat and mounting it in Ringer's solution on a glass slide with cover slip.

D. Wrap frogs in newspaper before discarding in appropriate waste container. Clean and replace equipment.

E. *Answer the following questions:*

1. Compare the location and function of the ciliated tissue just studied in the frog with that found in the human body:

   _____

   _____

   _____

   _____

2. When ether anesthesia is used in surgery the postoperative patient often has large accumulations of mucus in the respiratory tract, and his respirations are quite noisy. What is one possible explanation for this?

   _____

   _____

   _____

II. *Mechanical Factors Related to Breathing.* (work with partner)

A. *Movements of the thorax.*

1. Observe the changes in circumference of your partner's thorax during normal and forced breathing movements.

2. Place a tape measure around the chest at the axillary level. Hold tape in position while partner breathes. Record your measurements below:

Normal, quiet		Forced	
Inspiration	Expiration	Inspiration	Expiration

3. What muscles were used during normal, quiet inspiration?

_____

4. What additional muscles were called into action during forced inspiration?

_____

_____

5. Normal expiration is primarily a matter of _____

_____

but forced expiration involves the contraction of the _____

_____ muscles.

6. Examine the respiratory model consisting of bell jar fitted with rubber stopper, glass "Y" tubing, two balloons, and rubber sheeting.

   a. Pull down on the rubber sheet, and then allow it to return to its original position. Note results:

   _____

   _____

   b. How does this compare with muscle action in the human body? _____

   _____

c. Compare model with human system:

Model   *Human Counterpart*

glass tubing _____

balloons   _____

rubber sheeting_____

walls of bell jar_____

B. *Pressure changes related to movements of thorax.*

With the help of the respiratory model, text and reference books, complete the following statements:

1. Atmospheric pressure refers to _____

    _____

2. The outer surface of the lung is protected from this atmospheric pressure by

    _____

3. Intrapulmonic pressure refers to _____

    _____

4. Intrapleural or intrathoracic pressure refers to _____

    _____

    _____

5. During inspiration, muscular contractions cause the thorax to _____

    in size. This, in turn, causes a _____

    in the intrathoracic and intrapulmonic pressures. Since the pressure within the lung is now

    _____ than atmospheric pressure, air moves

    _____

6. During normal expiration the relaxation of muscles, combined with gravity, causes the thorax to

    _____ in size. this in turn, aided by elastic recoil of the **lungs, causes**

    an _____

    in the antrathoracic and intrapulmonic pressures. Since the pressure within the lung is now

    than atmospheric pressure, air moves _____

    _____

7. If the intrapleural pressure should ever become equal to atmospheric pressure what would happen to the lung?

_____

_____

_____

C. *Capacity of the lungs.*

1. What is meant by vital capacity? _____

_____

_____

2. Measure your own vital capacity by breathing into a spirometer. Record the largest volume of air expired in any one of three trials. _____ Liters.

3. Tidal air volume refers to _____

_____

4. Inspiratory reserve volume is the _____

_____

5. Expiratory reserve volume is the _____

_____

6. The volume of air remaining after the deepest possible expiration is called _____ Can it ever be removed from the lungs? Explain.

_____

_____

7. What is meant by minimal air volume? _____

_____

D. *Respiratory sounds.*

1. Place a stethoscope at various places on your partner's chest, and listen to the breath sounds during normal breathing, forced breathing, coughing, etc.

2. Listen in particular for rustling sounds which are due to the sudden expansion of alveoli during inspiration.

(Note: omit this exercise if clothing cannot be removed.)

III. *Chemical Factors Related to Breathing.*
(Have partner count and record your respirations below):

A. *Variations in respiratory rate.*

1. _____ per minute, while sitting quietly.

2. _____ per minute, after running in place for 1½ minutes.

3. Explain briefly, the reason for this change in rate:

    _____

    _____

4. Breathe deeply and vigorously, with mouth open, for 1½ to 2 minutes. Is there any great urge to breathe immediately following this period of hyperventilation? Explain:

    _____

    _____

    _____

    _____

5. Hold a paper bag firmly over your nose and open mouth while breathing deeply and vigorously for 2 minutes. Is there any great urge to breathe immediately following this period of hyperventilation? Explain:

    _____

    _____

    _____

    _____

B. *Voluntary control of breathing.*

1. How long can you hold your breath after a normal inspiration? _____seconds

2. After vigorous deep breathing for 2 minutes? _____ seconds.

3. After vigorous deep breathing into a paper bag for 2 minutes? _____seconds

4. Briefly, summarize and explain your results: _____

    _____

    _____

5. Would it be possible to commit suicide by holding your breath? Why? _____

    _____

IV. *The Exchange of Gases.*

   A. External respiration involves the exchange of gases between the _____ and the

   _____

   Internal respiration refers to the exchange of gases between the _____ and the

   _____

   B. *Diagram of alveoli and capillary network.*

     1. With blue pencil, lightly shade vessels containing blood low in oxygen.

     2. Use red pencil to shade those vessels containing blood that is well oxygenated.

     3. Add arrows to indicate the movement of oxygen and carbon dioxide between alveoli and capillary network.

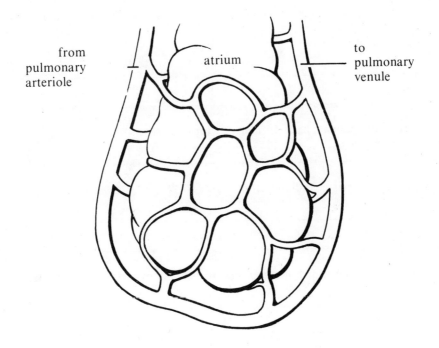

Figure 51.

V. *Application to Practical Situations.*

    A. List some of the factors responsible for initiating the first breath of a newborn baby: _____

    _____

    _____

    B. What is the respiratory distress syndrome? What may cause this disorder? _____

    _____

    _____

    C. Mrs. R. has tuberculosis of her right lung. In an effort to rest this lung and promote recovery, the doctor injects a measured amount of air into the pleural cavity.

        1. This procedure is called artificial _____

        2. How does it rest the lung? _____

        _____

        3. Is it permanent? Why? _____

        _____

        4. Crushing of the right phrenic nerve is another procedure that may be used to rest lung tissue. Explain briefly:

        _____

        _____

    D. Why do asthmatic patients have difficulty in breathing?

    _____

        1. Why does epinephrine (adrenalin) provide relief? _____

        _____

        _____

        2. Generally speaking, which muscles tend to become over-developed in these patients? _____

        _____

E. Briefly, explain the following terms:

1. hypoxia _____

2. dyspnea _____

3. orthopnea _____

4. apnea _____

   _____

5. asphyxia _____ _____

6. Cheyne-Stokes respiration _____ _____

   _____

7. atelectasis _____

   _____

F. What factors are involved in the increased work of breathing in patients with restrictive and obstructive lung disorders?

   _____

   _____ _____

   _____ _____

LABORATORY 20

The Digestive System

Digestive Tube and Accessory Organs

Demonstration Supplies:
 prepared microscope slides
 dissected embalmed cat or freshly killed animal
 fresh sheep liver with gall bladder and duodenum

Student Supplies:
 tongue depressors; flashlight
 dissectible human torso model
 human skulls; liver model
 anatomical wall charts
 text and reference books

I. *Demonstration of Digestive System in Cat.*

An anesthetized or freshly killed animal may be used if available. Otherwise, the dissected embalmed cat is quite adequate. The instructor may point out the major organs of digestion and show the interior of each when possible.

This demonstration is best done in small groups; therefore, students should proceed with the following work while waiting to see the animal.

II. *Diagram of the Digestive Tube and Accessory Organs.*

Turn to the next page and label the diagram as indicated. *Print* at the end of each line.

III. *The Mouth and Accessory Organs.*

Examine your partner's mouth. Use a wooden tongue depressor as necessary to visualize the following structures:

A. *Hard and soft palates.*

 1. The _____ and _____ bones form the hard palate which serves to separate

 the mouth from the _____

 2. The _____ is a membranous projection hanging from the lower margin of the soft

 palate.

The Digestive Tube and Accessory Organs.

anal canal
appendix
ascending colon
cecum
descending colon
duodenum
esophagus
hepatic flexure
ileum
jejunum

liver
pancreas
parotid gland
rectum
splenic flexure
sigmoid colon
stomach
sublingual gland
submandibular gland
transverse colon

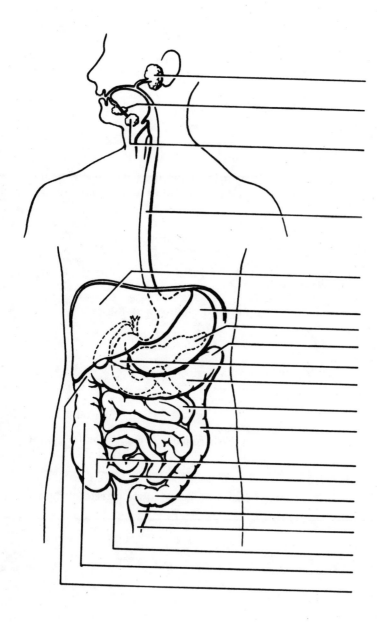

Figure 52.

B. *Palatine pillars and tonsils.*

  1. The anterior pillars are known also as the _____ arch; the posterior pillars

  are known as the _____ arch.

  2. If possible, note the palatine tonsils. They are found _____ the anterior

  and posterior pillars on each side.

C. *The tongue.*

  1. Note the small conical elevations studding the surface of the tongue. These are called ____

  _____

  2. The _____ is a prominent fold of membrane which connects the tongue

  with the floor of the mouth in the midline.

  3. Review the nerve supply to the tongue for

    a. general sensation: _____

    b. taste, anterior 2/3: _____

    c. taste, posterior 1/3: _____

    d. movement: _____

  4. What are the functions of the tongue? _____

  _____

  _____

D. *The teeth.*

  1. Examine your partner's teeth while the jaws are open wide. Identify: incisors, canines, premolars,
  molars. Compare with teeth in the prepared human skull.

  2. Are any "wisdom teeth" present? _____ What is the anatomical name

  for these teeth? _____

  3. That part of a tooth which can easily be seen is called the _____ ; the _____

  fits into a socket in the _____ process of the bone.

4. Examine teeth with jaws closed and lips pulled back.

    a. Which teeth meet? _____

    b. What is their function? _____

    c. Which teeth overlap? _____

    d. What is their function and what is the value of the overlapping? _____

        _____

5. Briefly, what is the correct way to brush your teeth? _____

_____

6. Deciduous teeth are commonly referred to as the _____

_____ teeth. They total _____ in number.

7. There are _____ teeth in the permanent set.

8. Label diagrams of the teeth on the following page.

E. *The salivary glands.*

1. With the aid of a tongue depressor examine the inner aspect of your partner's cheek on a level with the upper second molar. Note the small but prominent papilla in each cheek.

    Each papilla contains an opening for Stensen's duct which transports saliva from the

    _____ glands.

2. Examine floor of mouth under the tongue. Note small papillae lateral to the frenulum.

    These papillae contain the openings for Wharton's ducts which convey saliva from the

    _____ glands located near the inner surface of the angle of the mandible.

3. The _____ glands lie beneath the mucous membrane in the floor of the

mouth and have several small ducts opening under the tongue.

IV. *Pharynx and Esophagus.*

A. Examine the dissectible human torso model, wall charts, and reference books. Answer the following questions:

1. During the act of swallowing, the soft palate is drawn up to close the _____ and

prevent food from entering the nasal cavity.

a. *Diagram of the Permanent Teeth.*

Use colored pencils for identification:

blue — canines
brown — incisors
green — molars (tricuspids)
red — premolars (bicuspids)

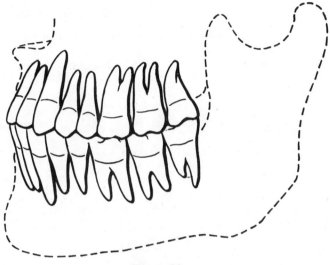

Figure 53.

b. *Vertical Section of a Tooth.*

alveolar process   enamel
blood vessels, nerves   gingiva (gum)
cementum   neck
crown   pulp cavity
dentin   root

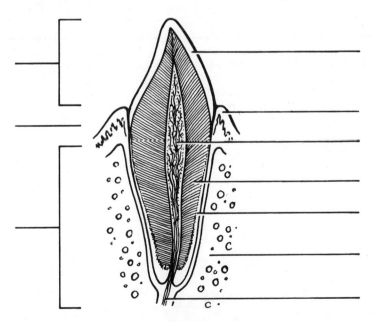

Figure 54.

2. As food leaves the mouth it passes through the fauces into the _____ pharynx, next it enters the _____ pharynx, and finally it goes into the _____ which opens into the stomach.

3. The _____ shunts food away from the larynx during swallowing and helps to prevent choking.

4. When passing a tube into a patient's stomach it might be helpful to remember that the average length of the pharynx is about _____ inches, and the esophagus about _____ inches.

The terminal portion of the esophagus lies below the diaphragm and at the level of the _____ process of the sternum.

V. *The Stomach.*

A. Examine viscera in the dissectible human torso model. Study wall charts and reference books while answering the following questions:

B. The stomach is located in the _____ and _____ regions of the abdomen.

C. One important fold of peritoneum is the _____ which attaches the intestines to the posterior abdominal wall; another fold is the _____ which connects the greater curvature of the stomach and the transverse colon.

D. What are the gastric rugae and what is their function? _____

_____

_____

E. The combined secretion of the various glands in the stomach is called _____,

and it contains _____

_____

F. What are the functions of the stomach? _____

_____

_____

G. *Longitudinal Section of Stomach.*

Print at the end of each label line:

cardiac orifice   lesser curvature
duodenum   pyloric sphincter
esophagus   pylorus
fundus   rugae
greater curvature

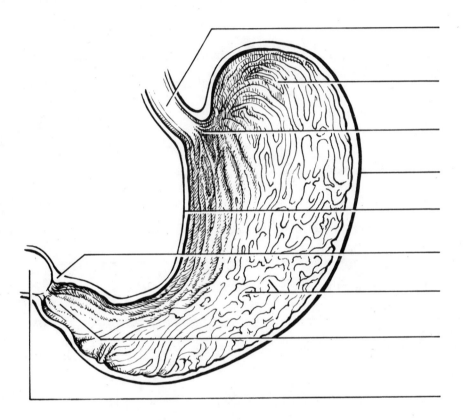

Figure 55.

VI. *The Intestinal Tract.*

A. Continue your study of models, charts, and reference books while answering the following questions:

B. The first 10 inches of small intestine is called the _____, the next division is known as the _____, and the distal or terminal portion is the

_____

C. What structures serve to increase the surface area for absorption in the small intestine? _____

_____

D. Peyer's patches are _____

that may be the seat of local inflammation and ulceration particularly in tuberculosis of the intestine.

E. The combined secretion of various glands in the lining of the small intestine is called

_____, and it contains the following enzymes:

_____

F. What are the functions of the small intestine? _____

_____

_____

G. The ileum opens into the _____ which is the first portion of the large intestine.

At this point a muscular sphincter called the _____ valve prevents the backflow of fecal

material.

H. What and where is the vermiform appendix? _____

_____

I. List, in sequence, the four subdivisions of the colon: _____

_____

J. What are the functions of the large intestine? _____

_____

_____

VII. *The Liver and Pancreas.*

A. Demonstration of fresh liver and gallbladder (with the duodenum, when possible).

Examine this specimen carefully and compare it with the model. Identify: lobes, blood vessels, and bile ducts. Squeeze the gallbladder and note appearance of bile in the duodenum which has been cut open with scissors.

B. In addition to peritoneum, the liver is covered by a fibrous tissue known as _____

_____

C. The liver produces a combined secretion-excretion, known as _____, which is stored

temporarily in the _____ after leaving the liver.

D. The _____ is the nutrient blood vessel of the liver, but the

_____ conveys blood that is rich in absorbed food from the gastrointestinal

tract.

   1. Blood from both incoming vessels passes into vascular channels which are called _____

   2. How do these channels differ from regular capillaries? _____

      _____

      _____

   3. Blood ultimately leaves the liver by way of _____

E. The biliary apparatus begins as _____ which unite with larger and larger ducts

until finally one main duct, the _____ , leaves the liver. The _____ duct from the

gallbladder unites with the duct from the liver to form the _____ bile duct which opens into

the duodenum at the _____

F. Summarize the major functions of the liver.

      _____

      _____

      _____

      _____

      _____

      _____

      _____

G. The pancreas is another important accessory organ. Its head rests in the curve of the _____

and the tail is in contact with the _____

H. The pancreas is both an endocrine and an exocrine gland.

   1. Insulin is the endocrine secretion that is essential for _____

      _____ , while glucagon has the effect of _____

      _____

2. The exocrine secretion of the pancreas is called _____, and it is conveyed to the duodenum through _____. The following enzymes are found in this secretion: _____

_____

I. *Diagram of the Biliary and Pancreatic Ducts.* (liver retracted upward)

Print at the end of each label line:

common bile duct	hepatic duct
cystic duct	liver
duodenal papilla	pancreatic duct
duodenum	pancreas
gallbladder	stomach

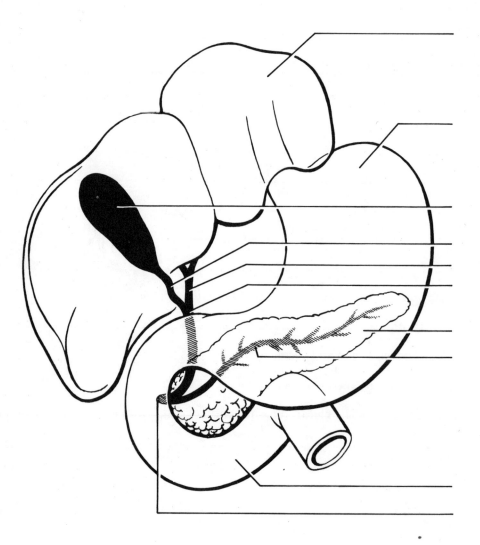

Figure 56.

VIII. *Study of Prepared Microscope Slides.* (optional)

With the aid of text and reference books examine the following slides:

A. The tongue.

    1. Locate: papillae, taste buds, muscle tissue

    2. Sketch one papilla showing general structure.

B. The small intestine.

    1. Locate: circular folds, villi, glands, goblet cells.

    2. Make a rough sketch and label it.

C. The liver, Golgi stain.

    Note the bile capillaries and ducts.

D. The liver, glycogen stain.

    1. Note areas of blue stain. This indicates glycogen.

    2. What is glycogen and why is it important? _____

_____

_____

IX. *Application to Practical Situations.*

A. In tongue-tied individuals the _____ is abnormally short.

B. Baby Ray was born with a cleft palate and must be fed with a medicine dropper.

    1. What is cleft palate? _____

_____

    2. Why can't Baby Ray take his milk from a bottle? _____

_____

_____

C. Peritonitis may follow rupture of the appendix.

    1. What is peritonitis? _____

_____

2. Why does peritonitis usually follow ruptured appendix? _____

_____

_____

D. An enema tube should be inserted for a distance of approximately _____ inches since the length of the anal canal is about _____ inches.

E. Why should children receive regular dental care even though their deciduous teeth are soon lost?

_____

_____

F. Briefly, explain the following terms:

1. gingivitis _____

2. dental caries _____

3. parotitis _____

4. gastrectomy _____

5. colitis _____

6. gastroenterostomy _____

_____

7. pyloric stenosis _____

8. cirrhosis _____

LABORATORY 21

Digestion, Absorption, and Metabolism

Demonstration Supplies:
    6 test tubes
    400 ml. beaker
    thermometer
    blue litmus solution
    5% pancreatin solution (dissolve
      5 Gm. of powdered pancreatin in
      100 ml. of 0.5% Na_2CO_3 solution)

    Benedict's solution
    1% starch solution
    sweet cream
    hard boiled egg

Student Supplies:
    crackers; paper cups; stethoscopes
    cooked starch solution (1%) with 100 cc. HNO_3 per liter added.
      (about 50 cc. per student needed)
    dilute iodine solution; test tubes and rack
    potassium iodide tablets, 0.325 Gm. (gr. v); reference books

I. *Demonstration of Enzyme Activity.*

    (Note: Demonstrations A and B may be set up in advance by the instructor, or they may be done as a student activity. Demonstration C must be set up in advance.)

  A. *Digestion of emulsified fat*

    1. Preparation.

      a. Pour 10 ml. of sweet cream into a test tube labeled #1.

      b. Add blue litmus solution to the cream until it has a distinctly blue color. Mix thoroughly.

      c. Pour half of this mixture into a test tube labeled #2.

      d. Add 2 ml. of 5% pancreatin solution to tube #1. Mix thoroughly.

      e. If the color of the mixtures in the two tubes is not approximately the same, add enough 0.5% Na_2CO_3 to tube #2 to make the colors match.

      f. Place both tubes in a beaker of water at 40°C. Allow to stand for at least one hour. Observe the color of the mixture in each tube.

    2. What was the color of tube #1? _____

      Of tube #2? _____

    3. What compounds caused the color change in test tube #1? _____

    4. Although it is not visible in the test tube, another end product of fat digestion is _____

5. The pancreatic enzyme responsible for the digestion of fat is _____

B. *Digestion of starch.*

    1. Preparation.

        a. Pour 5 ml. of 1% starch solution into a test tube labeled #3.

        b. Add 2 ml. of 5% pancreatin solution. Mix thoroughly.

        c. Place in a beaker of water at 40°C. for about 30 minutes.

        d. Take 1 ml. of the mixture in tube #3, add 5 ml. of Benedict's solution, and then heat gently until a color change is evident.

        e. As a control, place 1 ml. of a 1% starch solution in a test tube labeled #4, add 5 ml. of Benedict's solution, and heat gently for the same length of time as you did #3.

    2. What was the color change seen in tube #3? _____

    3. How does this color change prove that there has been some digestion of starch? _____

        _____

    4. What pancreatic enzyme is responsible for the breakdown of starch? _____

C. *Digestion of protein.*

    1. Preparation.

        a. In each of two test tubes, labeled #5 and #6, place a piece of cooked egg white (the pieces should be identical in size).

        b. To tube #5 add 10 ml. of distilled water.

        c. To tube #6 add 10 ml. of 5% pancreatin solution.

        d. Allow these tubes to stand in a warm place for several hours, then refrigerate until time for class.

        e. Observe the pieces of egg white in each tube.

    2. What evidence was there that the egg white in tube #6 was being digested? _____

        _____

    3. What was the appearance of the egg white in tube #5? _____

    4. What pancreatic enzymes are responsible for beginning the digestion of protein? _____

        _____

II. *The Absorption and Excretion of a Salt.*

Students may work in pairs or in groups with one person serving as the subject. Best results will be obtained in those sections scheduled for late morning or late afternoon.

A. Pour about 5 cc. of thin, cooked starch solution into each of 10 test tubes.

B. Add one drop of dilute iodine solution to the first test tube. Do NOT shake the tube. What color

change is noted? _____

C. Expectorate some saliva into the second test tube. Is there any color change? _____ . This

serves as the control tube to show that there is normally no appreciable amount of iodine in the saliva.

D. Place a tablet of potassium iodide, 0.325 Gm., well back on your tongue and swallow quickly with at least 250 cc. of water. Rinse mouth thoroughly after taking the tablet. Note the time.

E. Two minutes after taking the tablet, expectorate into the third test tube. *Do NOT shake the tube.*

F. Continue to collect saliva every two minutes until you have used all of your prepared tubes.

G. How many minutes elapsed before you were able to detect the presence of iodine in your saliva?

_____ minutes.

H. Compare your results with those of other groups. What was the time range? _____ to

_____ minutes.

I. Why is there more rapid absorption of this salt when it is taken before meals? _____

_____

J. List, in sequence, those structures through which the potassium iodide passed from the moment it left

the mouth until it was excreted in the saliva:

mouth ⟶ _____

_____

_____

_____

_____

III. *Digestive Processes.*

    A. The cooking of foods often plays an important part in the process of rendering them available to the human organism.

        1. List some advantages of cooking: _____

           _____

        2. Are there any disadvantages? _____

           _____

    B. What is meant by mechanical digestion in the digestive tube? _____

      _____

    C. Chemical digestion, on the other hand, involves a process of _____

      _____

    D. What effect does strong emotion, such as rage or pain, have upon the digestive processes? ____

      _____

IV. *Activity in the Mouth.*

    A. Dry the inside of your mouth with clean gauze or paper towel, and then attempt to eat a dry soda cracker.

        1. What important mechanical processes of digestion are illustrated by this exercise? _____

           _____

        2. Could you taste the cracker while it was dry? _____

        3. How does the location of taste buds affect ability to taste undissolved food? _____

           _____

    B. What stimuli promote the flow of saliva? _____

      _____

    C. Chemical digestion in the mouth is dependent upon the presence of _____ . This enzyme

      acts upon _____ and converts it to _____

    D. The term _____ is used in reference to a mass of food that is ready to be swallowed.

    E. What is meant by the term "absorption?" _____

      _____

    F. Why is there no absorption of food in the mouth? _____

      _____

V. *Deglutition.*

  A. Drink some water and note the action of your tongue. Also hold one hand on your throat during the swallowing process and feel the related movements.

  B. Attempt to swallow a mouthful of cracker while bending over a chair or stool so that your head is below the level of your stomach. Are you successful? _____

  1. During the first stage of deglutition your _____ pushed the bolus through the fauces.

  2. During the second stage:

    a. The soft palate was elevated to close the _____

    b. As the larynx moved upward and forward, the free edge of the _____ was moved downward, thus helping to block off the lower respiratory tract.

    c. Constrictor muscles of the _____ then passed the bolus into your _____

  3. During the third stage, the bolus moved along the esophagus by means of _____ until it finally passed into the _____

  C. Place a stethoscope over the xiphoid process of your partner's sternum while several large gulps of water are swallowed. What sounds do you hear? _____

  _____

VI. *Activity in the Stomach.* (Use text and reference books)

  A. *Mechanical processes:*

  1. How does the musculature of the stomach differ from that in the esophagus? _____

  _____

  2. Describe briefly the movements of the stomach:

  _____

  _____

  3. As digestion proceeds, small amounts of food are passed through the _____

  whenever the first part of the duodenum is empty. This semiliquid, acid food is called ____

  _____

  B. *Chemical processes:*

  1. List some of the stimuli that promote the secretion of gastric fluid: _____

  _____

2. The _____ nerves are the efferent pathways for reflex secretion of gastric fluid.

3. The pH of gastric juice averages 2 to 5 due to the presence of _____ acid. This acid serves to activate _____ by converting it to _____ ; the enzyme that attacks amino acid linkages in the center of _____ molecules.

4. A fat-splitting enzyme called _____ is secreted in very small amounts. Because its principal activity is on _____, converting them to _____ and _____ , it has very little effect on the large fat globules that make up the bulk of fat in the diet.

C. Absorption in the stomach is limited to _____

_____

VII. *Activity in the Small Intestine.*

A. *Mechanical processes:*

1. Briefly, describe the three types of movement which occur in the small intestine:

_____

_____

_____

2. Integration of intestinal activities is related to the extrinsic nerve supply of the walls and sphincters.

a. Stimulation of the vagus nerve promotes _____

_____

b. Thoracolumbar (sympathetic) nerves when stimulated cause _____

_____

B. *The role of bile.*

1. When food enters the duodenum, a hormone called _____ is produced by cells in the mucosa. This hormone then is carried in the bloodstream to the _____ , where it promotes the contraction of smooth muscle and causes concentrated _____ to be emptied into the duodenum.

2. Bile contains no enzymes, but the bile salts are essential for _____

C. *The pancreatic juice.*

1. A preliminary secretion of pancreatic juice is initiated by _____ nerve impulses, but

   hormonal regulation is of greater importance. The stimulus for liberation of these hormones is

   _____

2. The first hormone, _____, causes the pancreas to produce a watery juice that is rich in

   _____

3. The second hormone, _____, causes the pancreas to produce a thick secretion that is

   rich in _____

4. Why must the pancreatic proteases be secreted in inactive forms? _____

   _____

5. Pancreatic amylase attacks _____, converting then to _____ .

6. Pancreatic lipase splits fats into _____ and _____ .

D. *The intestinal juice* (succus entericus)

1. The flow of this juice is initiated by _____

   _____

2. Intestinal juice contains several enzymes called _____ which complete the

   digestion of proteins to _____

3. Carbohydrate digestion is completed as follows:

   sucrase splits _____ to _____

   maltase splits _____ to _____

   lactase splits _____ to _____

4. The enzyme enterokinase serves to activate _____

E. Explain briefly why the greatest amount of absorption takes place in the small intestine: _____

   _____

   _____

   _____

   _____

F. The absorbable end products of digestion are:

_____ from carbohydrates

_____ from proteins

_____ from fats

G. *Diagram of Circulation in an Intestinal Villus.*

Print at the end of each label line:

    arteriole
    blood capillary network
    lymph capillary
    venule

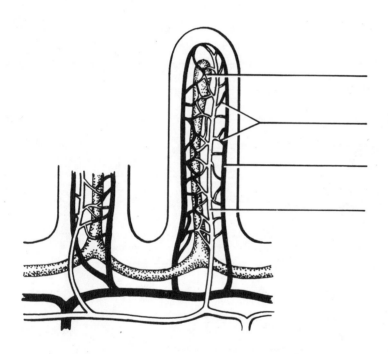

Figure 57.

H. *Pathways of absorption.*

1. The end products of fat digestion, _____ and _____ , enter the intestinal

epithelial cells, where they are resynthesized to _____ ; these droplets then enter the

highly permeable _____ capillaries.

2. Simple sugars and amino acids are absorbed by _____

_____ mechanism; they enter _____ capillaries.

3. Sodium and chloride are absorbed by _____ . As these and other inorganic salts are absorbed, the intestinal fluid becomes _____ tonic, thus water will be absorbed by the process of _____ .

VIII. *Activity in the Large Intestine.*

A. What is the significance of the "mass movements" that occur in the large intestine? _____

_____

Why do they represent the "gastrocolic reflex?" _____

_____

B. The secretion produced by the glands of the large intestine is lacking in enzymes and consists primarily of _____

C. There is great bacterial activity in this area, and it is responsible for:

1. putrefaction of _____ residues.

2. fermentation of _____ residues.

3. synthesis of vitamin _____

D. Absorption in the large intestine is limited to _____

_____

E. List the major constituents of feces: _____

_____

F. Which of the above is responsible for the normal color of feces? _____

_____

G. Defecation refers to _____

_____

1. What stimulus arouses the desire to defecate? _____

_____

2. Defecation is accomplished by relaxation of the _____ , plus contractions of smooth muscle in the walls of the _____

3. This evacuation is aided by voluntary contraction of the skeletal muscles of the _____ and the _____

4. Why is it important to empty the rectum as soon as there is a desire to defecate? _____

_____

_____

IX. *Metabolism.*

   A. Metabolism refers to _____

        _____

       1. Anabolism is the metabolic phase in which _____

           _____

       2. Catabolism, on the other hand, is the phase in which _____

           _____

           _____

       3. All metabolic processes are dependent on _____, the organic catalysts that are present in every cell. If even one of these chemical compounds is missing or malfunctioning, the effects may be extremely serious; such individuals are said to have an inborn _____

           _____

   B. The energy that is required for all metabolic activities is obtained from food by a process of biologic _____; much of the energy so released is stored temporarily in the high-energy phosphate bonds of _____ until needed by the body.

       1. The potential energy value of any given food is referred to as its _____ or heat value.

       2. _____ refers to the energy output required to keep an individual alive, i.e., to maintain body temperature, respiration and circulation.

   C. The main purpose of all carbohydrate metabolism is the ultimate provision of _____; however, glucose that is not needed at the moment may be converted to _____ and to _____ for temporary storage.

   D. Adipose tissue represents the storage form of lipids and, as such, provides _____

        _____

   E. The beta-oxidation of fatty acids results in the formation of a number of molecules of _____ which then enter the _____ cycle where they are oxidized with the liberation of _____ .

   F. The primary function of protein is to supply the amino acids necessary for _____

        _____

1. Amino acids in excess of the body needs are _____

    in the liver to form ammonia and keto acids; the ammonia is converted to _____

    _____ and excreted in the urine while the keto acids are converted to

    _____ ; if not needed for energy or for repair, the latter substances are

    converted to _____

2. Nitrogen balance refers to _____

    _____

X. *Body temperature.*

A. In healthy individuals the amount of body heat produced varies with _____

    _____

B. Most body heat is lost through the _____ , but some loss also occurs through _____

    _____

C. The temperature control center is in the _____

    1. In a cold environment the first line of defense against a change in body temperature is

        _____ ; the second line of defense includes _____

        _____

    2. In a hot environment, the first line of defense against a change in body temperature is

        _____ , the second line is _____ .

XI. *Application to Practical Situations.*

A. Unpleasant or painful procedures should not be carried out just before or just after mealtimes.

    Why? _____

B. In carcinoma of the rectum the rectum is removed surgically and a colostomy is provided.

    1. What is a colostomy? _____

        _____

    2. Why is a colostomy in the descending colon more easily controlled than one in the ascending

        colon? _____

        _____

        _____

C. Cholecystitis and cholelithiasis may cause obstruction of the common bile duct. A jaundiced
    appearance and clay-colored stools are common symptoms.

    1. What is cholecystitis? _____

        _____

2. Why might cholelithiasis cause obstruction of the common bile duct? _____

_____

_____

3. What is jaundice and why is it present? _____

_____

_____

4. Why are the stools clay-colored? _____

_____

5. Persons with cholelithiasis follow a low-fat diet. Why? _____

_____

_____

D. Why are non-absorbable cathartic salts effective in relieving constipation? _____

_____

E. Mr. H. C. was admitted to the hospital in a state of ketosis and acidosis.

1. What is ketosis? _____

_____

2. The primary ketone body is _____ ; two substances derived from it are

_____ and _____

3. Why may ketosis be accompanied by acidosis? _____

_____

_____

_____

F. Mr. X and Mr. Y work in a steel mill where the external temperatures are extremely high. Both of

them collapsed and were taken to the Medical Department.

1. Mr. X was very pale and perspiring profusely. He complained of muscle cramps, vertigo, and

nausea. The doctor said that he was suffering from heat _____ ; the symptoms were

due to _____

2. Mr. Y was comatose; his skin was dry and flushed. The absence of sweat is a striking

characteristic of heat _____ ; death invariably occurs if the body temperature rises

above _____

XII. *Chart for Summary and Review Purposes.* (optional)

Digestive Agents	Substances Acted upon	Products Formed
salivary amylase		
gastric protease (pepsin)		
gastric lipase (?)		
bile		
pancreatic lipase		
pancreatic proteases		
pancreatic amylase		
enterokinase		
intestinal proteases (erepsin)		
maltase		
sucrase		
lactase		

LABORATORY 22

The Urinary System; Water, Electrolyte and Acid-Base Balance

Demonstration Supplies:
    dissectible human torso model; projection slides
    anatomical wall charts; motion picture film
    dissected embalmed cat; x-ray films of human kidneys (pyelogram)
    prepared microscope slides
    Grammercy indicator solution
    1 ml. graduated pipette
    10 ml. diluted phosphate buffer solution
    10 ml. 2% gelatin solution
    N/10 HCl; medicine dropper
    Grammercy pH indicator chart
    3 test tubes

Student Supplies:
    fresh sheep or pork kidneys (one for each 2 or 3 students);
    dissecting trays and instruments; magnifying glasses; toothpicks;
    (optional: litmus paper; nitrazene paper; urinometers)
    text and reference books

I. *Dissection of a Fresh Sheep Kidney.*

    A. *Preliminary examination.*

        1. Note the large amount of adipose tissue which surrounds the kidney. Look for the adrenal gland which may still be embedded in the fat. (Note: the perirenal fat usually is missing if pork kidneys are ordered.)

            What purpose is served by this perirenal fat? _____

            _____

        2. Beginning on the convex border, peel off the fat until you reach the concave border. Here you will meet resistance due to the presence of certain tubular structures entering and leaving the kidney.

            a. These tubular structures are the _____

                _____

            b. They enter and leave the kidney through the _____

                which is a slit on the concave surface.

        3. Carefully dissect the fat away from these structures for a distance of at least one inch from the kidney, then cut across with sharp scissors so that the lumen of each tube may more easily be located.

        4. The _____ is the structure that conveys urine away from the kidney; it is

            lined with _____ membrane. Place a toothpick in the lumen of this tube.

5. If possible, identify the renal blood vessels. The mouth of the artery will be open, but the thin-walled vein will be collapsed.

B. If pork kidneys are being used, it is best to open the kidney by way of the ureter and renal pelvis. After observing the calyces and papillae, proceed as with sheep kidney.

C. With a sharp knife, make a longitudinal incision around the *convex* border of the kidney. Continue this incision down to the white basin, but DO NOT CUT ALL THE WAY THROUGH.

You should now be able to lay the kidney down much like an open book.

1. The toothpick which was inserted into the ureter should now be visible. What part of the kidney did it enter? _____

2. The calyces in pork kidneys more closely resemble their human counterparts than do those of sheep kidneys. Compare with textbook pictures.

3. Locate the renal cortex (reddish brown, granular in appearance).

4. What microscopic structures are located in the cortex?

_____

5. Locate the renal medulla (red, striated appearance).

6. The medulla is composed of cone-shaped structures called _____ . Their bases are directed toward the _____ , and their apices project centrally toward the _____ of the kidney.

7. The _____ are extensions of cortical substance which dip between adjacent pyramids and support blood vessels.

8. With a blunt probe gently press on the base of a pyramid and continue forward to the apex. Note the drops of urine which appear.

9. Repeat the above procedure, on another pyramid, while making your observations through a magnifying glass. Can you see the openings of collecting tubules?

D. Carefully dissect away the pelvic fat, and trace the course of renal blood vessels.

E. Wrap kidney in newspaper before discarding in appropriate waste container. Clean and replace equipment.

II. *Demonstration Units.*

A. Examine urinary organs in the dissectible human torso model and in the dissected embalmed cat. (The instructor may help with the cat.)

B. With the aid of text and reference books, examine the prepared microscope slides of the kidney. Identify the various parts of a nephron unit. Label the following diagrams.

The Urinary Organs and Related Structures.

Print at the end of each label line:

adrenal gland
aorta
bladder (distended)
common iliac a. and v.
external iliac a. and v.

internal iliac a. and v.
left kidney
renal a. and v.
left ureter
vena cava

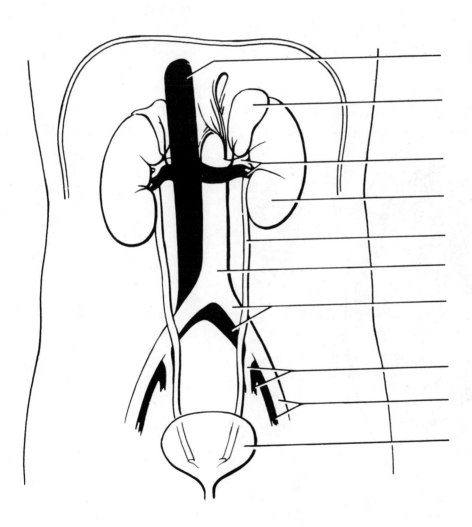

Figure 58.

The Kidney and Its Blood Supply.

1. Print at the end of each label line:

   afferent arteriole
   arcuate a. and v.
   calyx
   cortex
   interlobar a. and v.
   interlobular a. and v.
   nephron unit (greatly enlarged)

   renal a. and v.
   renal column (Bertini)
   renal papilla
   renal pelvis
   renal pyramid
   ureter

2. With colored pencils, lightly shade the following: (optional)

   ureter and calyces — yellow
   renal pyramids — red
   renal cortex — brown

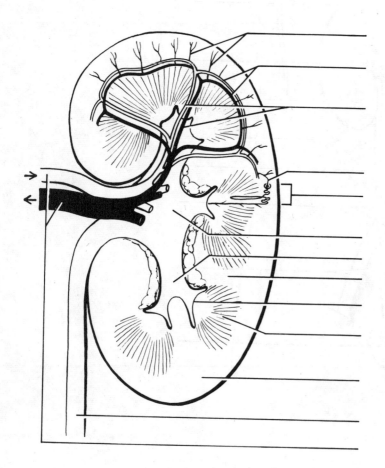

Figure 59.

A Nephron Unit and Related Structures

1. Print at the end of each label line:

   afferent arteriole
   arcuate, a. and v.
   Bowman's capsule
   peritubular capillary
   collecting tubule
   distal convoluted tubule

   efferent arteriole
   glomerulus
   interlobular a. and v.
   loop of Henle
   papilla
   proximal convoluted tubule

2. With colored pencils, lightly shade the following: (optional)

   arteries   — red
   veins   — blue
   renal tubule   — green
   collecting tubule — brown

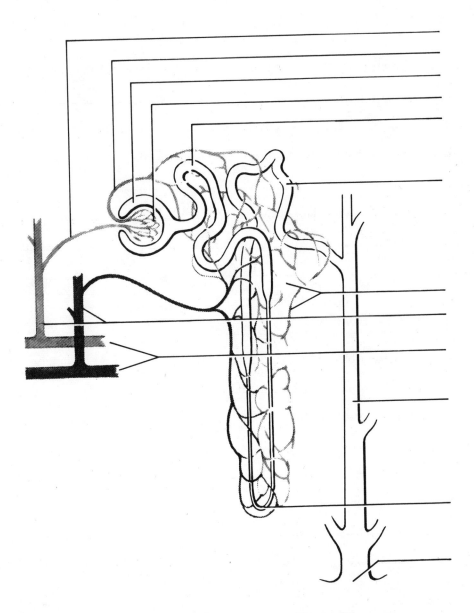

Figure 60.

C. A motion picture film, projection slides, and pyelogram studies may be shown when available.

III. *Answer the following questions:*

A. Describe briefly, the location and relative position of the kidneys: _____

_____

B. The work of the kidney centers on _____

_____

_____

C. The _____ are tubes that convey urine from the kidneys to the _____ .

D. Each kidney is composed of functional units called _____ .

E. What anatomic feature contributes to the maintenance of a relatively high blood pressure in the normal glomerulus?

_____

F. How is the glomerular blood pressure related to the preliminary phase of urine formation?

_____

_____

G. What substances are normally unable to pass through the glomerular walls? _____

_____

H. The final phase of urine formation takes place in the _____

and involves the processes of _____

_____

I. What is the function of ADH (antidiuretic hormone)? _____

_____

J. What is the function of the urinary bladder? _____

_____

K. Emptying of the urinary bladder involves contraction of the _____ muscle.

L. The urethra is the tube that _____

_____

M. Distinguish between the male and the female urethra:

a. function: _____

_____

b. curvature: _____

_____

c. length: _____

N. Micturition refers to _____

_____

O. What is the *trigone* of the bladder? _____

_____

IV. *Characteristics of Urine.* (optional)

A. Urine specimens may be obtained from the laboratory, or students may provide their own.

B. Describe the color of your specimen: _____

C. Is it clear or cloudy in appearance: _____

D. Reaction and pH

1. Test with litmus paper: _____ reaction.

2. Test with nitrazene paper: _____ pH.

E. Specific gravity

1. Pour urine into cylinder of urinometer until it is about 3/4 full.

2. Place urinometer in cylinder and allow it to come to rest. It should float freely, not touching sides or bottom of cylinder.

3. At eye level, observe the urinometer scale and locate the line that is in contact with the lower level

of the meniscus of the urine. What is the reading?

_____

4. Specific gravity may vary from _____ to _____ in normal kidneys.

5. What does the specific gravity indicate in regard to kidney function? _____

_____

V. *Fluid and Electrolyte Balance.*

A. Exogenous water is _____ while endogenous water is _____

B. What are the four channels through which water is eliminated from the body? _____

_____

C. A study of heat and water loss from the skin.

1. What is insensible perspiration? _____

2. What is sensible perspiration? _____

3. Every day we lose from 300 cc. to 700 cc. of water through insensible perspiration. By use of certain tables and simple arithmetic you can determine the amount of water loss in your own body.

   a. Your *total surface area* will show the extent of surface from which heat may be lost. Refer to the Du Bois Body Surface Chart (see page 235) to determine the area for your height and weight.

   What is it? _____ square meters.

   b. How many *Calories* will you produce *per hour per square meter* of body surface? (see Calorie Table, page 235.)

   The number is _____ Calories.

   c. You can now determine your *total heat production.*

   _____ Calories per hour.

   _____ Calories per 24 hours.

   d. Of this total heat production in 24 hours only *14.5%* is related to the evaporation of insensible perspiration. In your case that would be _____ Calories.

   e. Each gram of water evaporated absorbs 0.58 Calories of heat. Therefore, your daily insensible cutaneous loss of water would be _____ cc.

D. What is one major difference between intracellular fluid and extracellular fluid? _____

   1. The predominant extracellular ions are _____

   2. The predominant intracellular ions are _____

   3. The term "electrical neutrality" means _____

   _____

   4. Sodium and potassium are excreted in _____ , _____ and _____ , but the _____ are the organs that assume the major role in maintaining sodium and potassium balance.

E. Calcium and phosphate are absorbed from _____

   1. The rate of calcium absorbtion is largely determined by the _____

   2. One regulating factor is _____ hormone which promotes increased absorption as well as a shift of calcium into the ECF from storehouses in _____ tissue.

   3. A hormone that opposes an increase in calcium levels in the ECF is _____ .

   4. Phosphate absorption depends largely on _____

VI. *Acid-Base Balance.*

   A. The term "acid-base balance" refers to _____

_____

     1. The normal pH of the blood is _____

     2. Most body fluids are alkaline in reaction, but two important exceptions are _____

   B. Chemical buffers. (Demonstration)

     1. Place 0.5 ml. of Grammercy Indicator Solution in a clean test tube, add 10 ml. of distilled water, then shake the tube to mix contents.

       a. Read the pH by comparison with the chart.

       b. Add 1 drop of N/10 HCl and shake the tube.

       c. Read the pH then set the tube aside as a reference.

     2. Place 0.5 ml. of Grammercy Indicator Solution in a second test tube, add 10 ml. of diluted phosphate buffer solution, then shake the tube to mix contents.

       a. Read the pH.

       b. Add N/10 HCl drop by drop, shaking the tube after the addition of each drop, until the color is the same as that obtained in 1.b. above.

       c. Record the number of drops of N/10 HCl required.

     3. Place 0.5 ml. of Grammercy Indicator Solution in a third test tube, add 10 ml. of 2% gelatin solution, then shake tube to mix contents.

       a. Read the pH.

       b. Add N/10 HCl drop by drop until the color is the same as that obtained in 1.b. above.

       c. Record the number of drops of N/10 HCl used.

     4. Is protein effective as a buffering agent? _____

       Is it as effective as phosphates? _____

     5. Why did the pH finally change even with the phosphate buffer? _____

_____

     6. In the carbonate buffer system, the ratio of $NaHCO_3$ to H_2CO_3 is _____. This system operates very effectively in the _____, preventing major changes in the pH of that fluid.

C. Respiratory and renal mechanisms.

1. The carbonate system buffers many of the relatively strong acids produced by metabolic activities, forming _____ which is a weak acid.

2. Rising blood levels of the above acid have what effect on the respiratory rate? _____ How does this help to get rid of acid? _____

_____

3. The renal tubular cells help to maintain acid-base balance by secreting varying quantities of _____ ions in exchange for _____ ions. The tubular cells also can form _____ which is secreted into the tubular lumens so that additional _____ ions may be returned to the bloodstream.

VII. *Application to Practical Situations.*

A. In acute nephritis, the kidneys temporarily stop producing urine; hemodialysis may be necessary.

1. What is nephritis? _____

2. Hemodialysis refers to _____

_____

_____

3. Briefly, what is the value of this procedure. _____

_____

_____

B. After a major abdominal operation there may be difficulty in voiding, and catheterization of the bladder may be required.

1. What is catheterization? _____

_____

2. Why is sterile equipment used when inserting catheters? _____

_____

3. What is cystitis? _____

_____

C. The following terms are used in reference to abnormal constituents in urine. Identify each.

1. glycosuria: _____

2. hematuria: _____

_____

3. albuminuria: _____

4. pyuria: _____

5. casts: _____

D. What is the general function of a diuretic drug? _____

    _____

    Why are mercurial diuretics effective? _____

    _____

E. Identify the following terms:

    1. glomerulonephritis: _____

        _____

    2. polyuria: _____

    3. anuria: _____

    4. pyelitis: _____

        _____

F. Drink a glass of water. List, in sequence, the structures through which this water travels from the time
    it leaves the mouth until part of it is excreted as urine:

    mouth _____

    _____

    _____

    _____

    _____

    _____

G. Edema may be the result of an increased effective _____ pressure or a decreased

    _____ pressure.

H. Electrolyte imbalances.

    1. Hyponatremia means _____

        It usually results from _____

        What is the effect on cardiac output? _____

        On blood pH? _____ , on the skin? _____

204

2. Hyperkalemia is a manifestation of _____

   It may develop when _____

   _____

   What organs are most critically affected? _____

   _____

3. Hypocalcemia means _____

   What effect does this have on the nervous system? _____

   _____ , on skeletal muscle? _____

I. Metabolic acidosis is caused by _____ ,

   but respiratory acidosis is due to _____

LABORATORY 23

The Endocrine Glands

Demonstration Supplies:

    dissected embalmed cat
    preserved human specimens
    dissectible human torso model
    motion picture film
    projection slides (optional)

Student Supplies:

    text and reference books
    anatomical wall charts

I. *Demonstrations.*

    A. The instructor will meet with small groups and point out the endocrine glands as seen in the embalmed cat.

    B. Examine the human torso model. Note the location and position of adult endocrine glands.

    C. A motion picture film may be shown when available.

    D. Examine reference books which illustrate the results of hyper- and hypofunction of the various endocrine glands.

II. *Introductory Questions.*

    A. What is a gland? _____

    _____

    B. Distinguish between the following and give one example of each:

        1. exocrine glands: _____

        _____

        2. endocrine glands: _____

        _____

    C. What are hormones? _____

    _____

    D. Generally speaking, the activities of the endocrine system supplement those of the nervous system in that they aid in _____

E. Why is cyclic AMP called the "second messenger"?

_____

What is the "first messenger"? _____

III. *The Pituitary Gland.* (hypophysis)

A. This gland is located _____

_____

_____

B. It is composed of two parts:

1. The anterior lobe or _____

because of its glandular nature.

2. The posterior lobe or _____

because it is derived from the nervous system.

C. There are at least 6 physiologically important hormones that have been identified as originating from the anterior lobe of this "master gland."

Indicate, briefly, the functions of the following anterior pituitary hormones:

1. growth hormone or somatotropin: _____

_____

2. adrenocorticotropin (ACTH): _____

_____

3. thyrotropin (TSH): _____

_____

4. prolactin, lactogenic, or luteotropin (LTH): _____

_____

_____

5. gonadotropins: _____

a. follicle-stimulating hormone (FSH): _____

_____

_____

b. luteinizing hormone (LH): _____

_____

_____

D. The posterior pituitary stores 2 hormones that were originally produced by special cell bodies in the

_____

What functions are performed by the following:

1. antidiuretic hormone (ADH) (vasopressin): _____

    _____

    _____

2. oxytocin: _____

    _____

IV. *The Thyroid Gland.*

  A. This gland is located _____

     _____

     and it consists of two lateral _____ which

     are united by a strip called _____

  B. The thyroid hormones, thyroxin and triiodothyronine, are concerned with _____

     _____

     _____

  C. The hormone thyrocalcitonin serves to _____

     _____

V. *The Parathyroid Glands.*

  A. Locate and describe these glands: _____

     _____

     _____

  B. Parathyroid hormone is essential for _____

     _____

     _____

     _____

VI. *The Adrenal Glands.*

   A. Where are these glands located? _____

      _____

   B. Each gland is composed of an inner portion, which is called the _____,

      and an outer portion, the _____.

   C. The adrenal medulla secretes a mixture of 2 hormones:

      1. _____is important in emergency

         situations since it reinforces and prolongs the activities of the _____

         division of the autonomic system.

      2. _____ does not affect carbohydrate

         metabolism but does cause widespread peripheral vasoconstriction resulting in _____

         peripheral resistance and _____ both

         the diastolic and the systolic blood pressures.

   D. Most hormones of the adrenal cortex are essential for life as they regulate many important functions. Briefly, summarize the action of the following:

      1. mineralocraticoids (e.g. aldosterone): _____

         _____

      2. glucocorticoids (e.g. cortisol): _____

         _____

      3. The production of aldosterone is controlled by _____

         _____

      4. Production of the glucocorticoids is primarily under the control of _____

VII. *Gonadal Hormones.*

   A. Estrogen is produced by _____

      1. Its production is controlled by _____

         from the anterior pituitary.

      2. Estrogen is concerned with _____

         _____

         _____

B. Progesterone is produced by _____

    1. Its production is controlled by _____

      hormone from the anterior pituitary.

    2. Progesterone is responsible for _____

      _____

      _____

      _____

C. Testosterone is produced by _____

   and it is essential for _____

   _____

   _____

VIII. *The Pancreas.* (a heterocrine gland)

A. The beta cells of the _____ secrete

   insulin which is essential for _____

   _____

B. Glucagon, produced by the alpha cells, serves to _____

   _____

IX. *The Prostaglandins.*

A. These hormonelike substances are formed within _____

B. Functionally, they serve as _____

   as they can interact with _____ ,

   turning it on or off.

X. *Application to Practical Situations.*

A. A friend of yours has just been admitted to the hospital for treatment of diabetes mellitus. She is very upset by dietary restrictions and the hypodermic injections of insulin. How would you answer her questions?

    1. "Why can't I have all the spaghetti I want? After all, that isn't sugar!"

      _____

      _____

      _____

2. "Why can't you just put that insulin in a glass and let me drink it?"

_____

_____

3. "How soon will the doctor cure me so that I can get out of the hospital?"

_____

_____

_____

B. In exophthalmic goiter (Graves' disease), a subtotal thyroidectomy is done if treatment with drugs is ineffective.

1. Exophthalmic goiter is a form of _____

2. The primary cause of this disease is thought to be an overproduction of thyrotropic hormone by

the _____

gland; overactivity of the _____ gland is a secondary factor.

3. List a few of the more prominent symptoms that one might expect to observe in this patient

_____

_____

_____

4. What did the doctor hope to accomplish by the use of a drug such as propylthiouracil? ___

_____

5. What is a subtotal thyroidectomy? _____

_____

C. Identify the endocrine imbalance that is associated with each of the following:

1. Diabetes insipidus _____

_____

2. Acromegaly _____

_____

3. Tetany _____

4. Addison's disease _____

5. Cretinism _____

_____

6. Myxedema _____

_____

7. Gigantism _____

_____

8. Cushing's syndrome _____

_____

LABORATORY 24

The Reproductive System

Part A—**Organs of Reproduction**

Demonstration Supplies:

    male and female dissected, embalmed cats
    preserved and fresh human specimens
    prepared microscope slides
    projection slides

Student Supplies:

    anatomical wall charts
    text and reference books
    dissectible human torso model

I. *Demonstrations.*

Note:While awaiting your turn at the demonstration areas, study wall charts and reference books. Proceed with sections II and III on the following pages.

A. *Organs of reproduction in the cat.*

The instructor will help you in identifying prominent reproductive structures as seen in male and female cats.

B. *Human torso model.*

Examine the reproductive organs in the dissectible model.

C. *Prepared microscope slides.*

1. Examine the sections of ovarian tissue.
   Locate: ovarian follicle and corpus luteum.

2. Examine the section of testicular tissue.
   Locate: seminiferous tubules, ciliated epithelium.

3. Examine the slide of human spermatozoa. Note that each sperm has a head, middle piece, and tail piece.

D. *Preserved and fresh human organs (when available).*

e. g. uterus, tubes, ovaries; scrotum and testes

II. *Female Reproductive Organs.*

A. *Internal structures.*

1. The female sex cells are called _____ ,

   and each develops within _____ near the

   surface of the ovary; development of these structures is related to _____

   hormone from the anterior pituitary.

2. In addition to housing sex cells, follicular cells also secrete a hormone called _____

_____

3. What is meant by ovulation? _____

_____

4. Following ovulation, the cells of the follicle enlarge and develop yellow pigment. This structure is

called the _____ ,

and it secretes a hormone called _____

5. Where do oocytes go after leaving the follicles? _____

_____

6. What are the functions of the uterus? _____

_____

_____

7. Briefly, identify the following:

    a. myometrium _____

    b. cul-de-sac or pouch of Douglas _____

    _____

    _____

    c. endometrium _____

    d. broad ligament _____

    _____

    e. uterosacral ligament _____

    _____

8. The vagina is located posterior to _____

and anterior to _____ . It serves as the

_____

_____

B. *External structures.*

1. The area between the pubis, anteriorly, the tip of the coccyx, posteriorly, and the thighs laterally

    is called the _____ ; the area between vagina and

    anal canal is called the _____

2. The external genitalia may also be referred to as the _____ .

3. What is the mons pubis? _____

_____

4. The _____ is a small body of erectile tissue

covered by the prepuce, a small hood of skin formed by the _____

5. Briefly, describe the location and function of the mammary glands: _____

_____

_____

C. *Label the following diagrams:*

Print at the end of each label line.

1. *External Genitalia of the Female.*   *Lithotomy Position.*

    anus   obstetrical perineum
    clitoris   (perineal body)
    labia majora   prepuce
    labia minora   urethral orifice
    mons pubis   vaginal orifice
                           vestibule

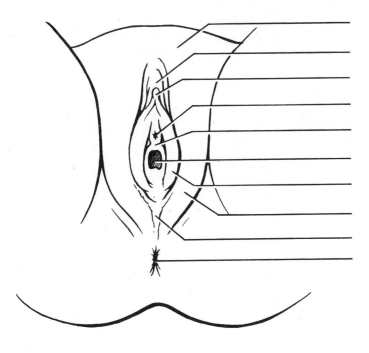

Figure 61.

2. *Female Pelvis.   Midsagittal Section.*

body
cervix
clitoris
cul-de-sac (Douglas)
fundus
labia majora
labia minora

posterior fornix
rectum
urinary bladder
urethra
vagina

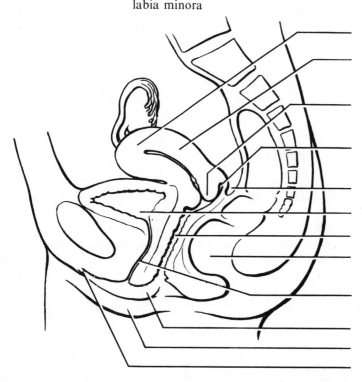

Figure 62.

3. *The Mammary Gland and Related Structures.*

adipose tissue
areola
glandular tissue
intercostal muscle
lactiferous duct

nipple
pectoralis major
second rib
sixth rib

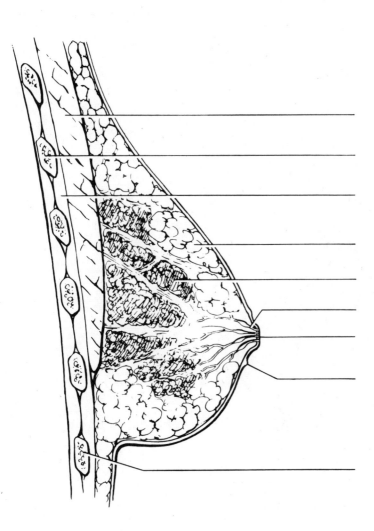

Figure 63.

4. *Female Reproductive Organs. Coronal Plane.*

(highly diagrammatic)

broad ligament
cervix
corpus luteum
endometrium
fimbriated end of tube
fornix
fundus
greater vestibular gland

labia majora
labia minora
ligament of ovary
mature follicle
myometrium
ovary
uterine tube
vagina

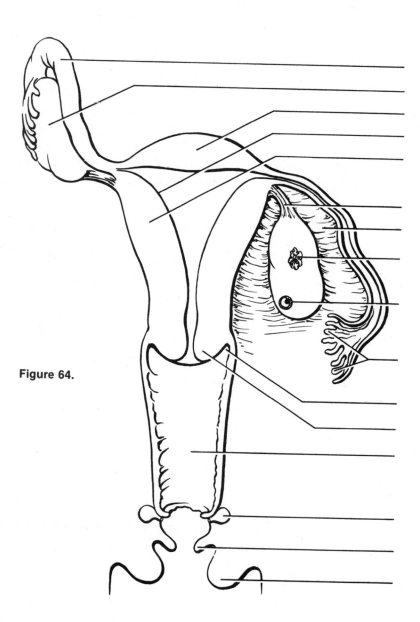

Figure 64.

III. *Male Reproductive Organs.*

A. *The testes.*

1. Male sex cells are called _____ ,

   and they are formed in the _____

   of the testes.

2. In early fetal life the testes are found in the _____ ,

   but a short time before birth they usually descent through the _____

   and enter the _____ .

3. Failure of the testes to descend from their original position will result in sterility.

   Explain: _____

   _____

   _____

4. In addition to the formation of spermatozoa, the testes also produce a hormone,

   called _____ ,

   which is responsible for _____

   _____

B. *Excretory ducts.*

1. As spermatozoa leave the testes they pass into the _____

   where the maturation process continues.

2. Briefly, describe the course of the ductus deferens:

   _____

   _____

   _____

3. Sperm pass from each ductus deferens into the two _____

   ducts which are surrounded by the _____ gland.

4. The final portion of the seminal pathway involves the urethra which is subdivided into three

   parts:

220

a. The _____ portion

which receives seminal fluid from the _____ ducts.

b. The _____ portion

which passes through the pelvic floor of the body.

c. The _____ portion

which lies within the penis.

C. *Accessory structures.*

1. As sperm are delivered to the ejaculatory ducts the _____

add their secretion, which provides _____

_____

2. Next, the _____ gland

adds its secretion, which also is _____ in

reaction and serves to _____

3. Ducts of the _____ glands

open into the cavenous urethra; they produce a _____ secretion

which probably aids in _____

_____

4. Semen is _____

_____

a. In each cubic centimeter of average semen there are approximately _____

_____ mature sperm.

b. How many sperm are required to fertilize the oocyte? _____

c. Why does nature provide so many? _____

_____

d. Mature spermatozoa have _____ chromosomes.

5. The _____ ,

or male organ of copulation, enables semen to be deposited in the vagina near the cervix of the uterus.

D. *Label the following diagrams.*

Print at the end of each label line.

The Male Duct System.   Sagittal Section

(highly diagrammatic)

bulbo-urethral gland
cavernous urethra
ductus deferens (vas)
ejculatory duct
epididymis
glans penis
membranous urethra
prepuce

prostate
prostatic urethra
scrotum
seminal vesicle
symphysis pubis
testis
ureter
urinary bladder

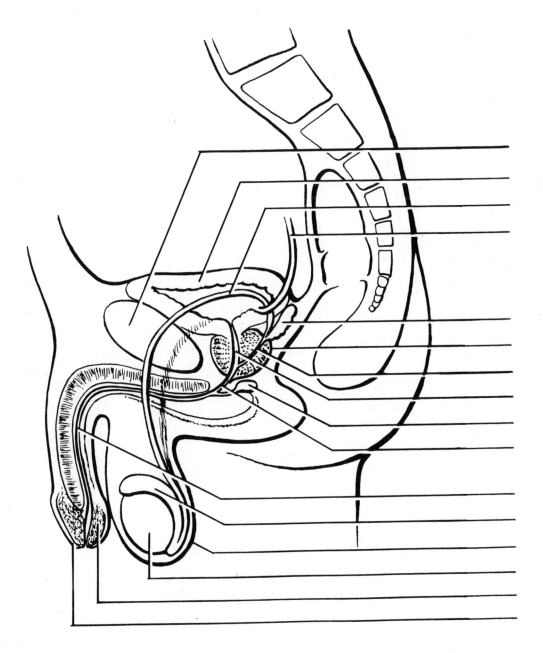

Figure 65.

2. *Testes and Epididymis.* *Sagittal Section.*

    ductus deferens (vas)
    ductus epididymis
    lobules
    rete testis
    seminiferous tubules

Note: Indicate, with arrows, the direction of movement of sperm from the time of their formation until ready to leave the body.

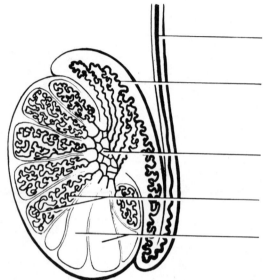

Figure 66.

IV.   *Application to Practical Situations.*

A. Explain the following in simple terms:

1. endometritis _____

    _____

2. cervicitis _____

    _____

3. rectocele _____

    _____

4. pruritus vulvae _____

    _____

5. leiomyoma _____

    _____

6. salpingitis _____

7. orchitis _____

B. When doing a pelvic examination, the doctor introduces two fingers into the vagina so that he can palpate the _____ of the uterus. His other hand presses down on the lower abdominal wall and palpates the _____ of the uterus.

   1. The normal position of the uterus is _____

   _____

   _____

   2. Retroversion of the uterus refers to _____

   _____

C. A "pubic-perineal shave prep." is done prior to repair of a cystocele.

   1. What specific structures in this area are covered with hair? _____

   _____

   2. What is a cystocele? _____

   _____

D. Briefly, what is meant by

   1. dilatation and curettage (D and C): _____

   _____

   2. mastectomy: _____

   _____

   3. cesarean section: _____

   _____

   4. vaginal hysterectomy: _____

   _____

   5. circumcision: _____

   _____

   6. orchidectomy: _____

   _____

E. Mr. C. G. is to undergo a *trans-urethral resection* of an enlarged prostate gland (benign prostatic hypertrophy). His chief complaint is "much difficulty in voiding."

1. What is the anatomical basis for his chief complaint? _____

_____

_____

2. A transurethral resection means removal of the prostate gland through the _____

_____ with special

instruments that electrically cut away excess tissue.

Part B — Reproductive Processes

Demonstration Supplies:

> dissected embalmed cat: pregnant female
> fresh human placenta
> motion picture film;   projection slides

Student Supplies:

> reference books and charts
> preserved specimens (e.g. human fetus, fetal pig series)

I.   *Demonstrations.*   (by the instructor)

Note: While awaiting your turn at the demonstration areas, study the wall charts and reference books. Examine the preserved human fetus and series of fetal pigs. Answer questions in Sections II and III.

A.   *Pregnancy in the cat.*   (embalmed)

The feline uterus has two horns and is designed for multiple births; however, it is interesting to note position of kittens, arrangement of fetal membranes, umbilical cord, etc. Also note the displacement of other abdominal viscera by the enlarged uterus.

B.   *Human placenta.*   (fresh)

Observe the tremendous network of blood vessels, the chorion, amnion, and umbilical cord.

C.   Motion picture film and projection slides may be shown near the end of the period.

II.   *The Menstrual Cycle.*

A.   The average duration of one cycle is _____days, and

during this time there are typical changes in the functional layer of the _____

_____or uterine mucosa.

B.   The first day of the cycle is said to be the first day of _____

_____ even though this act is in reality

the final event of each cycle.

1.   Briefly, what is menstruation? _____

_____

2.   How long does the average menstrual flow last? _____

days.

3. Approximately how much blood is lost during an average menstrual period? _____ cc.

C. The preovulatory phase, also called the _____

phase, is associated with a rapidly growing _____

and the production of _____ hormone.

What effect does this hormone have on the endometrium?_____

_____

Two pituitary gonadotrophins influence this phase, the first being, _____ ,

and the second _____ .

D. On approximately the 14th, day of the cycle, _____

occurs and the oocyte, with its _____ chromosomes,

enters the _____ .

   1. This event marks the beginning of the _____ or

   _____ phase.

   2. A corpus luteum develops at the ovulatory site; it produces estrogen and a second ovarian

   hormone called _____ .

   3. The second hormone is responsible for _____

   _____

   _____

E. If the oocyte is *not* fertilized, the corpus luteum undergoes _____ ,

and the _____

begins again.

III. *Pregnancy and Parturition.*

A. Fertilization refers to _____

_____

   1. This union usually takes place in the _____ ;

   it results in a single cell which has _____ chromosomes.

   2. The most favorable period for fertilization occurs around the middle of the menstrual cycle when

   _____ takes place.

3. What happens to the corpus luteum when the ovum is fertilized? _____

_____

B. List, in sequence, all of the structures through which spermatozoa must pass from the time of their formation until an oocyte is fertilized: _____

_____

_____

C. As the fertilized ovum moves toward the uterus it begins to undergo rapid cell division by

_____

1. In about _____days after fertilization a berry-shaped group of cells (morula) reaches the uterus.

2. Implantation then occurs. Briefly, explain the meaning of this term: _____

_____

D. As cell division continues, the embryo grows rapidly in size and differentation of germ layers begins.

(*Optional*: Refer to reference books for a listing of these germ layers and the adult structures which arise from them).

E. Identify the following embryonic and fetal structures:

1. blastocyst _____

_____

2. chorion frondosum _____

_____

3. amnion _____

_____

4. amniotic fluid _____

_____

5. umbilical cord _____

_____

F. Identify the following maternal membranes (decidua) which are derived from the uterus:

1. decidua basalis: _____

_____

2. decidua capsularis: _____

_____

G. What are the functions of the placenta? _____

_____

_____

_____

H. *Schematic View of Fetal and Maternal Membranes.*

Label the following diagram as indicated:

amnion
amniotic fluid
chorion
chorion frondosum

decidua basalis
decidua capsularis
fetus
umbilical cord

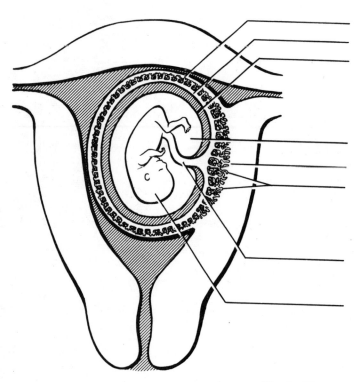

Figure 67.

I. *The intra-uterine or prenatal period lasts about* _____

weeks, or _____

lunar months.

   1. The time from the third week to the end of the eighth week is called

      the period of the _____

   2. The time from the ninth week until the baby is born is known as the

      period of the _____

J. Labor or parturition refers to _____

_____

State briefly the main event in each of the three stages of labor:

   1. First stage: _____

   2. Second stage:_____

   3. Third stage: _____

IV.   *Postnatal Life.*

  A. The neonatal period extends from _____ to the _____

  B. The period of infancy extends from _____

_____

to _____

  C. Childhood embraces the time from _____

until _____

  D. What is puberty? _____

_____

    1. Puberty occurs in girls at about _____ years,

       but in boys the average age is about _____ years.

    2. What secondary sex characteristics become manifest?

      a. in girls: _____

         _____

      b. in boys: _____

         _____

230

E. Puberty in the female terminates at menarche, a term that refers to _____
_____

F. Menopause refers to the _____
_____

　　1. The average age of onset of menopause is between _____ and
_____ years.

　　2. At this time the ovaries no longer _____
_____or
_____ and the period of
_____ is over; soon there is
atrophy of _____
_____

V. *Application to Practical Situations.*

A. During a discussion about childbirth, someone used the term "ectopic pregnancy."

　　1. Generally speaking, what is an ectopic pregnancy?
_____

　　2. Therefore, a tubal ectopic pregnancy would mean that _____
_____

　　3. Why do tubal pregnancies usually fail to reach full term? _____
_____
_____

　　4. Will future pregnancies be possible after a tubal pregnancy?
Explain: _____
_____
_____

B. Removal of both ovaries in a young woman will produce the symptoms of premature menopause.

　　1. Briefly, what are some symptoms of menopause? _____
_____
_____

2. Generally speaking, what role do the ovaries play in preventing the onset of menopause?

_____

_____

C. Mrs. V. B. has a large leiomyoma of the uterus. Her doctor has advised hysterectomy.

1. What is a hysterectomy? _____

2. Would pregnancy be possible following such a procedure? _____

3. Would she ovulate? _____ . Would she menstruate?

_____

    Explain: _____

_____

D. Identify the following terms:

1. amenorrhea _____

2. dysmenorrhea _____

3. sterility _____

4. lactation _____

_____

APPENDIX

I. *Complete List of Laboratory Equipment Used in This Manual.*

A. *Instruments and metal-ware:*

blood lancets (or needles)
flashlight
forceps, tissue
forceps, hemostatic
microscopes
needles, hypodermic
needles, teasing

pins, dressmaker
plain dividers
probes, blunt
ring stands and clamps
scalpels
scissors
sphygmomanometers

spirometer (vital capacity)
stethoscopes
stop watch
trays, dissecting, with wax
tuning forks
urinometer
water thermometer

B. *Glassware:*

beakers
capillary tubing
funnels
glass tubing
graduated cylinders

hypodermic syringes
magnifying lenses
microscope slides
    and cover slips
stirring rods

test tubes
watch glasses

C. *Chemicals and drugs:*

acetic acid
acetone
amyl nitrite, aspirols
alcohol, ethyl
Benedict's solution
blue litmus solution
copper sulfate
egg albumin, powdered
ether (anesthetic)
gelatin
Grammercy Indicator
    solution

hydrochloric acid
iodine solution
methylene blue, 1%
mineral oil
phosphate buffer
    solution
nitric acid
nux vomica, Tr.
pancreatin (powdered)
potassium iodide (0.325 Gm.)
potassium permanganate
sodium carbonate

sodium chloride
sodium citrate (or oxalate)
spirits of ammonia
sucrose
starch
strychnine sulfate
urethane, ethyl
wood charcoal, powdered

D. *Miscellaneous items:*

cork boards, with 3/4″
    hole at one end
cotton, absorbent
cotton tipped applicators
dialyzing membrane
distilled water
filter paper

glass marking pencils
ice
lens paper
litmus paper
nitrazene paper
paper cups
paper bags, brown, #5

plain cork
rubber stopper, one-hole
tape measures
tongue depressors
toothpicks
Tallqvist hemoglobin scale
wooden racks for test tubes

234.

E. *Models:*

brain, dissectible
ear, dissectible
eye, dissectible
heart, dissectible

kidney
larynx
liver
human torso, dissectible

respiratory (bell jar with
  balloons and diaphragm)
spinal cord and nerves
  (X-section)

F. *Osseous materials:*

dissectible skull
fetal skull

skeleton, articulated
skeleton, disarticulated

sphenoid bone
temporal bone

G. *Preserved materials:*

doubly and triply injected cats, one male and two female
  (one pregnant and one nonpregnant), if possible
fetal pig series of embryos
human specimens, if possible, e. g. brain, heart, kidney,
  uterus, tubes, ovaries, fetuses, etc.

H. *Fresh materials:*

amoeba culture
brains, sheep
blood, human
  (optional)
bones, beef (longitudinal
  and transverse sections)
chicken egg (hard boiled)

chicken or rooster legs
crackers, soda
eyes, beef
frogs, small rana pipiens
hearts, sheep (or pork) and
  beef joints, beef
kidneys, sheep or pork

liver, sheep (with gallbladder
  and duodenum)
lungs, sheep, with trachea
  and larynx
placenta, human
prunes, dried
sweet cream

I. *Other visual aids:*

series of prepared microscope slides showing sections of
  various human tissues
Ciba collection of 2 x 2 color projection slides
motion picture films
slide projector
motion picture projector
X-ray view box and assorted films
series of anatomical wall charts
reference books, e.g. anatomical atlases

11. *Body Surface Chart and Calorie Production Table.*

(see p. 200) The Du Bois chart below determines body surface area. Connect the correct height (column 1) with the correct weight (column 2) using a straightedge. The straightedge intersects column 3 at the surface area. (Prepared by W. M. Boothby, R. B. Sandiford)

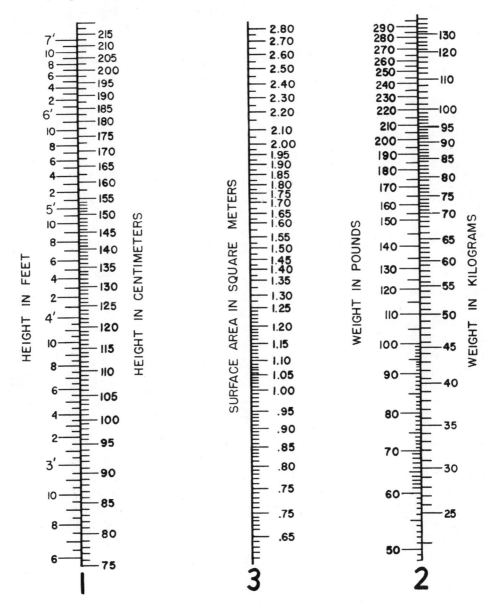

THE BOOTHBY AND SANDIFORD NORMAL STANDARDS

	CALORIES PER SQUARE METER PER HOUR				
AGE	MALES	FEMALES	AGE	MALES	FEMALES
5	(53.0)	(51.6)	20-24	41.0	36.9
6	52.7	50.7	25-29	40.3	36.6
7	52.0	50.7			
8	51.2	48.1	30-34	39.8	36.2
9	50.4	46.9	35-39	39.2	35.8
10	49.5	45.8	40-44	38.3	35.3
11	48.6	44.6	45-49	37.8	35.0
12	47.8	43.4			
13	47.1	42.0	50-54	37.2	34.5
14	46.2	41.0	55-59	36.6	34.1
15	45.3	39.6	60-64	36.0	33.8
16	44.7	38.5	65-69	35.3	33.4
17	43.7	37.4			
18	42.9	37.3	70-74	(34.8)	(32.8)
19	42.1	37.2	75-79	(34.2)	(32.3)

III. *Formulae for Laboratory Solutions.*

A. *Physiological Saline:* (human)

sodium chloride .... 9.0 Gm.
add distilled water to .... 1,000 cc.

B. *Ringer's Solution for Frogs:*

sodium chloride .... 6.5 Gm.
sodium bicarbonate .... 0.2 Gm.
calcium chloride .... 0.1 Gm.
potassium chloride .... 0.1 Gm.
add distilled water to .... 1,000 cc.

C. *Normal Saline for Frogs:*

sodium chloride .... 6.5 Gm.
add distilled water to .... 1,000 cc.

D. *Solutions for Fixing and Preserving Fresh Tissues:*

Jore's Solution #1:

sodium chloride powder	90 Gm.	or	18 Gm.
sodium chloride	50 Gm.	or	10 Gm.
sodium sulfate (crystalline)	110 Gm.	or	22 Gm.
chloral hydrate	200 Gm.	or	40 Gm.
formalin (10%)	300 cc.	or	60 cc.
add tap water to	10,000 cc.	or	2,000 cc.

Jore's Solution #2:

sodium bicarbonate (powder)	90 Gm.	or	18 Gm.
sodium chloride	50 Gm.	or	10 Gm.
sodium sulfate (crystalline)	110 Gm.	or	22 Gm.
chloral hydrate	100 Gm.	or	20 Gm.
formalin (10%)	50 cc.	or	10 cc.
add tap water to	10,000 cc.	or	2,000 cc.

Note: Wash the fresh specimen thoroughly in normal saline. Fix in solution #1 for one or two weeks or longer according to size of specimen. Transfer to solution #2 for final mounting.

E. *Urethane Solution:* (10%)

ethyl urethane .... 10.0 Gm.
add distilled water to .... 100 cc.

F. *N/10 HCl:*

add 4.2 cc. of hydrochloric acid to 400 cc. of distilled water; then add more water to a total of 500 cc.